A Field Guide to Forest Seeds

NIPA® GENX ELECTRONIC RESOURCES & SOLUTIONS P. LTD.
New Delhi-110 034

About the Authors

Ashwath M.N., Research Scholar, Department of Forest Biology and Tree Improvement College of Forestry, KAU, An enthusiastic scholar in the field of Forestry. Currently perusing Doctoral Degree in the field of Forest Biology and Tree Improvement. Awarded with a gold medal in M.Sc. (For. FBTI) from the University of Agricultural and Horticultural Sciences, Shivamogga. Qualified ASRB-NET and gained research and teaching experience. Honored with DST-Inspire fellowship for Doctoral research.

Dr. Sathish B.N. is currently working as Asst. Professor in College of Forestry, Ponnampet, Keladi Shivappa Nayaka University of Agricultural and Horticultural Sciences, Shivamogga. He has twelve years of experience in teaching, research and outreach activities in the field of Forestry. Before joining as Asst. Professor in the university, he worked in a project on 'Mapping of plant resources of Western Ghats' funded by DBT, New Delhi for five years. During this period he has extensively worked in the forests of Western Ghats. He has handled more than 10 research projects related to biodiversity conservation and livelihood enhancement. He has participated in various national and international conferences, seminar and symposia. he also published more than 50 research papers in peer reviewed national and international journals. He has published three books independently. He was a recipient best teacher award for the year 2020 by Keladi Shivappa Nayaka University of Agricultural and Horticultural Sciences, Shivamogga.

Dr. G.M. Devagiri is the doyen of tropical forestry. Currently he is working as Professor and Head, Dept. of Natural Resource Management and Dept. of Forest Products and Utilization, and University Head, Division of Forestry and Environmental Sciences, Keladi Shivappa Nayaka University of Agricultural and Horticultural Sciences, Shivamogga. He has more than 25 years of experience in teaching, research and outreach activities in different aspects of tropical forestry.

He has implemented several research programs in Western Ghats, India. Notably, DOS-DBT– Biodiversity characterization in south Karnataka, Vegetation carbon project under ISRO-IGBP program, Sentential Landscape project funded by ICRAF-CGIAR, Quantification of ecosystem services in Central Western Ghats, funded by Canadian Model Forest Network.

He has attended several national and international conferences, seminars and symposia. He has been actively associated with many national and international organizations. He has been serving as member of Western Ghats Eco-sensitive Zone Monitoring Committee, member of Kodagu Wildlife Conflict Mitigation Foundation, Govt. of Karnataka. He is serving as the President of Agricultural Sciences Forum of Kodagu in India.

Dr. Devagiri's interest in Forest Ecology, Remote sensing and GIS, Biodiversity conservation, Climate Science has placed him as one of those forestry scientist who lay hands on whisper of nature and bring in the science and aesthetic value of nature. He is an prolific writer and has published about 100 research papers in peer-reviewed national and International journals. He has also published many books and book chapters on Forestry and Environmental sciences.

Clara Manasa P.A. is a dedicated researcher in the field of Agroforestry and Silviculture. Completed B.Sc. and M.Sc. Forestry from the University of Agricultural and Horticultural Sciences, Shivamogga and currently pursuing PhD from the same University. Qualified ASRB-NET and has gained experience in teaching as Assistant Professor (CT).

A Field Guide to Forest Seeds

Ashwath M.N.
Department of Forest Biology and Tree Improvement
College of Forestry
Kerala Agricultural University
P.O., Vellanikkara, Thrissur, Kerala-680 654

Sathish B.N.
Department of Forest Products and Utilization
Department of Forest Biology and Tree Improvement
College of Forestry, Ponnampet
K.S.N. University of Agricultural and Horticultural Sciences
Shivamogga, Karnataka

G.M. Devagiri
Department of Natural Resource Management
and Forest Products and Utilization
College of Forestry, Ponnampet
K.S.N. University of Agricultural and Horticultural Sciences
Shivamogga, Karnataka

Clara Manasa P.A.
Department of Silviculture and Agroforestry
College of Forestry
K.S.N. University of Agricultural and Horticultural Sciences
Shivamogga, Karnataka

NIPA® GENX ELECTRONIC RESOURCES & SOLUTIONS P. LTD.
New Delhi-110 034

NIPA® GENX ELECTRONIC
RESOURCES & SOLUTIONS P. LTD.

101,103, Vikas Surya Plaza, CU Block
L.S.C.Market, Pitam Pura, New Delhi-110 034
Ph : +91 11 27341616, 27341717, 27341718
E-mail: newindiapublishingagency@gmail.com
Website: www.nipabooks.com

For customer assistance, please contact
Phone: + 91-11-27 34 17 17
Fax: + 91-11-27 34 16 16
E-Mail: feedbacks@nipabooks.com

ISBN: 978-81-19002-28-3

NIPA® also publishes books in a variety of electronic formats. Some content that appears in print may not be available in electronic books, and vice versa.

Composed and Designed by NIPA®.

Acknowledgement

The authors would like to convey our special thanks to the College of Forestry, Ponnampet, for giving an opportunity and encouragement. Special thanks to Dr. B. Gurdevsingh, Rtd. Scientist, IFGTB, and Karnataka Forest Department for all the support. It's our pleasure to reiterate our profound indebtedness and a deep sense of gratitude and heartfelt thanks to the whole forestry family who helped directly and indirectly.

Preface

Hobby to professional...!!

Seeds are incredible because they are smaller versions of the entire plant, concealing the entire tree structure under the skin and allowing them to thrive in favorable conditions. Seeds have been known to humans from the beginning of time, and they have played an important role from nomadic farming to precision farming. These are employed in a variety of ways, including food, decorative, cosmetics, medicine, pharmaceuticals, and so on, harnessing their physiochemical or biological qualities.

It made me to think about how a seed's form, size, and color varies from species to species. Being a forester instilled in me the habit of collecting the seeds of various tree species that I come across on field trips, outdoor exploration, excursions, and other activities. This passion prompted me to begin gathering basic information about seed and their handling, and I was able to come out with this compiled data as a ***"A Field Guide to Forest Seeds".***

The journey from collecting seeds to writing and gathering information was joyous, and it added to my understanding of how seeds struggle and adapt. Some are heavy as stone and cannot flow, while others are swimmers, some fly in the air, some taste sweet or bitter, some are naked or ingested in fruit pulp, and range in number from a single seed to a large number. All of the characters are examples of adaptation to nature's rules, which distinguishes them and adds variety.

Seed descriptions, including phenology, distribution, and basic seed handling techniques for more than 75 tree and plant species utilized in forestry programs, have been compiled and presented with seed photos in this book. We believe this gathered material will greatly benefit students, field personnel, and others interested in forest nursery activities. This is the first attempt as a book, and it is anticipated that more species will be exposed in the future. Comments are happily accepted for improvement of the content.

Authors

Contents

1

Introduction

The forest cover in most tropical countries has been under severe pressure and is diminishing at an alarming speed; consequently, they are experiencing severe biodiversity losses, reverting to a mature state at a slower rate with low resiliency. These tropical countries are aware of the danger causing due to the loss of forests, and large afforestation programmes have been launched by forest departments, private companies, and donor-supported projects. Nevertheless, plantation activities have rarely been able to keep pace with the deforestation rate, not to mention replacing what has been lost during the past. Plantation programme through afforestation activity is not only the matter of the forest department alone but also the combined responsibility of all the foresters. The appeal of forest planting has changed over the years following the broader demand spectrum.

To some extent, the forest plantations fail to create the diversity of plants and wildlife that has been lost. Hitherto, large-scale plantations are still necessary both directly to cater to the needs of industrial wood and help to minimize the pressure on the remaining natural forests. During the past decades, many government programmes have emphasized using local species, where exotic species formerly dominated in plantations. With the abovementioned more comprehensive planting activities, this trend has increased the demand for a much wider diversity of species, including local and little-known species. The selection and establishment of the species essentially depend on the plantation area, species selection, and plantation types combined with the determinants that influence seed demand and, therefore, the seed supply. Concurrently with seed supply, appropriate technology for many lesser-known tropical and subtropical species needs to be developed.

Generally, plants reproduce and multiply in two ways: asexually (vegetative) and sexually by forming seeds. Vegetative propagation has become increasingly accepted for reproducing trees with problematic seeds. Yet, seed propagation remains the primary method of propagation in Silviculture in the tropical and temperate regions. A seed is a living entity and is described as the plant's propagative material or generative part to reproduce a new generation. It is the dormant plant stage creating a link between two generations. However, the most accepted definition of seed is a matured ovule consisting of an embryonic plant and stored food, all surrounded by a protective coat. Seeds are considered unique compared to the vegetative propagules in natural and artificial regeneration because:

1. Seeds constitute unique genetic compositions resulting from mixing parental genetic material. The result is a genetic variation of the offspring, enhancing ecological adaptability.
2. Seeds are usually produced in large numbers and are readily available yearly or at longer intervals.
3. Seeds are (usually) small concentrated packages of plants-to-be, containing nutrients for the establishment of the plant and, excepting recalcitrant seeds, usually much more resistant to damage and environmental stress than vegetative propagules.
4. Seeds of several species can be stored for long periods under cold, dry conditions.

The knowledge of seed biology comprehends the development and physiology of seeds until they finally germinate or fail to do so. Hence, complete information on seed biology is critical for properly managing seed sources and handling seeds themselves.

2

Seed Handling and Description

Seed handling ensures the potential utilization of seed material. Collection and utilization of quality seeds are indispensable in producing quality planting material and better seedling establishment. After selecting the quality seed source, collecting good quality seeds physiologically and genetically is considered the first step in seed handling, followed by processing seeds, testing, storage, and pre-treatment for germination. The main goal of seed handling is to attain an excellent seed survival rate. The chances of seeds getting damaged during any stages of seed handling because of careless or rough approaches are very high.

From the regeneration point of view, the quantity of seed production per tree correspondingly has a vital role to play. If considered virtually, only one successful seed (or two if the species is dioecious) would serve to replace the parent tree(s); hence the production of seeds during the lifetime of a tree is absurd. A full-grown *Eucalyptus camaldulensis* tree may produce a million or more seeds per year and live further, producing seeds for over a century. Each year's production could help afforest several hundred hectares. On the contrary, several other species may produce a smaller number of seeds. However, it is never a limiting factor in natural regeneration as each seed has the potential to become a full-grown tree. Still, in nature, most of the seed production will succumb - from failed dispersal, predation, infestation, natural deterioration, germination failure, etc.

Seed processing

The fruit or seed processing objective is to achieve clean, pure seeds of high physiological quality (germinability) that can be stored and easily handled during succeeding processes, such as pre-treatment, transport, and sowing. Processing includes several handling procedures where applicability differs, e.g., according to fruit and seed type, condition of the fruits or seeds at the collection, and potential storage period. Processing can be assorted into the following seven procedures:

1. Pre-cleaning for fruit or seed lots containing larger debris, leaves, twigs, empty fruit parts, etc.

2. Pre-curing for fruits that must be after-ripened or where rapid desiccation hampers extraction.
3. Extraction, for species where the fruits are collected, only the seeds (and occasionally part of the fruit) are stored and sown.
4. Dewinging, for fruits and seeds with wings. It also includes the removal of dry appendices like spines, arils, and hairs.
5. Cleaning for fruits or seeds with impurities like fruit parts, leaves, twigs, empty seeds, foreign seeds, and chaff.
6. Grading for seed lots with considerable seed size or weight variation.
7. Adjustment of moisture content for seeds after the other procedures have a particular species for the expected storage period.

Seed processing usually follows the order mentioned above, but specific steps may be irrelevant and hence may exclude particular species or seed lots.

Seed storage and seed moisture content

Seed storage is the technique of preservation of viable seeds until they are sown. Seed viability and storage are greatly affected by the storage environment, moisture content, temperature, etc. Based on longevity, seeds are classified as **Microbiotic** - seeds with a life span not exceeding three years; **Mesobiotic** - seeds with life spans from 3 to 15 years; and **Macrobiotic** - seed with a life span of 15 to 100 or more years. The storage period also determines the appropriate moisture content of species.

The seeds can be classified as orthodox and recalcitrant based on the available moisture content. **Orthodox seeds** can be dried to a minimum moisture content of around five percent and stored at a low or sub-freezing temperature for long periods. These seeds are hard and germinate by absorbing the moisture from the atmosphere when kept in an open condition. They are dried and can be stored in tightly-sealed containers for extended periods. The storage life of orthodox seeds doubles for every 20°C decrease *(e.g., Leucaena leucocephala, Albizzia, Cassia, etc.*). **Recalcitrant seeds** are seeds that cannot be stored for a longer period. On storing for an extended period, seeds will lose viability and fail to germinate. These seeds/fruits are necessary to be retained moist. For interim storing, the seeds can be packed in absorbent affluence such as sawdust, peat moss, etc., and preserved under suitable ventilated conditions. Under the best conditions, the viability does not retain more than a few days to weeks *(e.g., Mangifera indica, Hopea, Azadirachta indica, Artocarpus, etc.*).

Dormancy and pre-treatment

Seed dormancy refers to the physiological state in which viable seeds fail to germinate even when provided with conditions typically favourable to germination, i.e., adequate moisture, appropriate temperature regime, a normal atmosphere, and in some cases light. Dormancy has evolved to avoid germination under conditions where seedling survival is likely to be less or sometimes termed as delayed germination. There are several levels of dormancy varying from very slight to very strong (deep). Under certain circumstances, these levels may change during the lifetime of the seed, usually as a response to external conditions.

Dormancy may be of several kinds, and sometimes, one seed might express more than one type of dormancy, and there is no universally adopted system. Harper (1977) classified dormancy according to development into three types, *viz*. 1) **innate dormancy** in which dormancy is present when the seeds are ready for dispersal, 2) **induced dormancy** in which dormancy develops as a response to external environmental factors, and 3) **enforced dormancy** in which germination is constrained because of external conditions. However, the latter group does not comply with the usual definition of dormancy since germination according to that definition, does not occur despite favorable germination conditions, not because of unfavourable conditions. Seeds that do not germinate because of external conditions are more correctly referred to as **'quiescent'** (Villiers 1972). Innate and induced dormancy are also referred to as primary and secondary dormancy, respectively.

Any cause of dormancy related to the embryo, e.g., immature development or chemical inhibitors located in the embryo, may collectively be called **endogenous or embryo dormancy.** Analogously, mechanical resistance, physical impermeability, inhibitors, or light sensitivity associated with the seed-coat are called **exogenous** or **seed-coat dormancy**. 'Seed-coat' is used broadly in any enclosing structure, e.g., endocarp or the entire pericarp.

Two or more dormancy types occur in the same seeds, called **'double dormancy'** or **'combined dormancy.'** Double or combined dormancy is, for example, found in fleshy fruits with chemical inhibitors combined with, e.g., a hard endocarp (**physical dormancy**) or immature embryos combined with other dormancy types.

These kinds of dormancy are naturally broken by continued exposure to moisture and light or saprophytic soil organisms. Seed may germinate after being exposed to digestive acids in the gut of animals and birds. Scarification is the process of breaking down dormancy. Large seeds may scarify by damaging the seed coat by cutting or scraping. The objective is not to remove the seed coat but to create tiny holes to allow water to get in. Care should be taken to avoid injury to the embryo. Hot water scarification is generally followed for the small and minute seeds by soaking in hot water. Some of the seeds will germinate after overnight soaking in cold water. Soaking seeds in acids is another way of breaking dormancy. Alternate wetting and drying, Cow dung treatment, Sandbag treatment, soaking in milk, cow urine treatment, Gibberellic acid, etc., are generally in practice to interrupt the dormancy.

Germination and seedling establishment

Germination demarcates the transition from the seed dependent on food sources from the mother plant to an independent plant capable of taking up nutrients and growing independently. Hence, germination also makes up the last link in the chain of seed handling processes. The usual germination pattern starts with radicle protrusion, followed by the elongation of the embryonic axis, developing into the main stem. The seedling stem is divided into the hypocotyl, and the epicotyl, the former being section below the cotyledons, and the latter is the section above the cotyledons. The hypocotyl elongation has two forms that divide the mode of germination and seedling development into two main distinctive groups, **hypogeal (or hypogeous)** and **epigeal (or epigeous)**. In hypogeal germination, the hypocotyl does not expand or expands only slightly. The cotyledons (and hence the seed) remain below the ground during germination and do not become photosynthetic (cryptocotylar). In epigeal germination, the hypocotyl expands and pushes the cotyledons above the ground, often together with the seed-coat and possible remaining endosperm. The first appearance of epigeal germination is the loop of the elongated hypocotyl above the ground. Some seeds germinate inside the fruit when attached to the parent plant; such germination is called **vivipary**. Vivipary is prevailing in mangrove trees of the family Rhizophoraceae and is a common feature in high arctic grasses.

The development of young seedlings after germination varies with species but is also strongly dependent on the environment. For tree seedlings, the environment of the young plant is quite different from the environment experienced by the adult, and accordingly, environmental requirements or tolerance show adaptations to juvenile life. For example, seedlings of humid forest trees are often sensitive to full light exposure and desiccation during their juvenile stage; pioneers often show a rapid initial growth rate to give them a competitive advantage over weeds. Some dry zone species have delayed height growth after seedling emergence while an extensive root system is formed. Such seedlings must be planted out despite their low height since root pruning is often detrimental.

Seed Problems - Limiting Factors in Tree Propagation and Afforestation

Problems with seed procurement and technology often limit the use of particular species. Nurseries may tend to raise species whose seeds are readily available and easy to grow from seeds since the additional work imposed by more problematic species does not pay. For several species, seed shortage, difficulties in the collection, short viability and/or extraction problems, dormancy, or other handling difficulties limit their use in plantation programmes. For example, collecting seeds from mature trees of *Alnus nepalensis* and *Grevillea robusta* is challenging because the seeds are borne on thin and inaccessible branches at a great height and are easily lost during collection.

Many tropical humid-zone species like khayas and dipterocarps have the dual problem of seed availability: short seed viability combined with long intervals between seed crops. In others, limitations in seed viability (whether caused by inert seed physiology or limitations in storage technology) make immediate sowing necessary, sometimes not optimal for seedling survival during the nursery stage. An example is *Azadirachta indica*, which sometimes must be kept as tiny seedlings in the nursery through thc dry season. Extraction and dormancy problems make seed handling of some species difficult, thereby limiting their use. Examples are *Pterocarpus* species and some species from fire-prone areas. Seed viability often restricts the distance seeds can be used away from their parent source. On the international scale, short-lived species are prone to deterioration during transport and transit, even with an efficient transport system. Such problems affect the introduction of a species into a new area and the exchange of germplasm between countries.

There is a need for a systematic explanation of the principles for handling forest seeds and solving particular problems associated with tropical forest seeds. The present field guide endeavoured to record the phonological activities by observation on local flora around the Kodagu district (falling under a humid tropical region), and the methods followed for processing and treatments recommended are simplified and given in a concise way to address by the foresters, seed collectors, nursery workers, and anyone concerned with the use of forest seeds.

Acacia auriculifomis Benth

Family : Fabaceae (Mimosoideae)

Common name : Australian wattle.

Distribution : The species is native to Savannas of Papua New Guinea and the northern areas of Australia; it is also introduced in Indonesia, Malaysia, India, Tanzania.

Seeds of A. auriculifomis

Habit : This is a moderate-sized evergreen tree. Leaf stalks are modified into flattened blades called phyllode.

Phenology : Tree flowers profusely and fruiting almost every year. The seed matures from January to March, and the best time for collection is around February to March.

Uses: The timber is used for local furniture making; it is a beneficial tree for reclaiming wasteland areas and conserving water and soil. The bark contains tannin, which produces good quality leather.

How to identify and Seed Handling: Pods are 1.2 cm wide, flat, hard, almost woody, and much twisted in irregular coils, initially green but turn brown on ripening. They are splitting open along the edges. Seeds of one-gram weight contain 40 to 45 seeds. Germination is moderate, generally above 50 percent. Seeds are viable for nearly two years. Seeds soaked in water for 24 to 36 hours before sowing gives better germination. Treating seeds for 40 minutes in 50% conc. H_2S0_4 gives 85 percent germination (Natarajan and Rai, 1988).

Acacia concinna Linn.

Family : Fabaceae (Leguminosae)

Common name : Shikakai, Soap-pod.

Distribution : It is commonly distributed in tropical rainforests of Southern Asia. India shares its habitat most commonly in evergreen, semi-evergreen, and sometimes in moist deciduous forests (Khanpara *et al*., 2012).

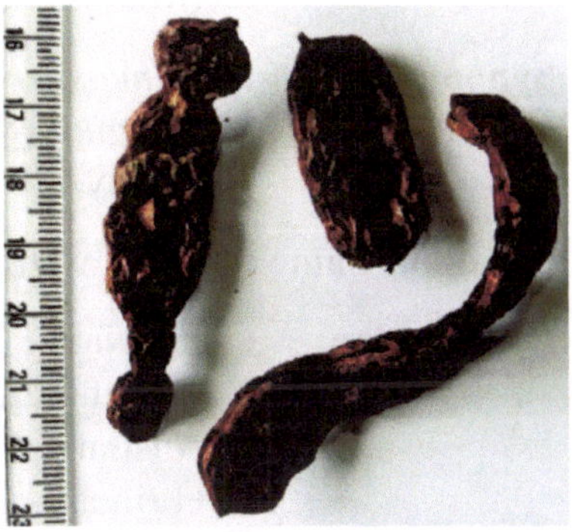

Pods of Shikakai

Habit : Evergreen, thorny woody climber and even found as a small shrub when failed to reach the host for climbing.

Phenology : Flowers throughout the year. Matured pods are produced from March to April and November to January.

Uses : Most commonly used medicinal plant, the fruits are used for washing hair, promoting hair growth, emetic, as an expectorant and purgative, and used in manufacturing body and facial creams.

How to identify and Seed Handling: Pods are long, flat, and stiff with 5 to 8 seeds. Seeds have a hard seed coat which avoids germination. Around 7000 to 10000 seeds weigh one kilogram. Seeds can be stored for one to two years in ambient conditions. Scarification is the better way to overcome dormancy and for easy germination. Germination is 40 to 90 percent, starting from seven days of sowing and completing after the 20th day.

Acacia mangium Willd.

Family : Fabaceae (Mimosoideae)

Synonyms : *Rancosperma mangium* (Willd.) Pedley

Common name : Black wattle.

Distribution : It is Native to northern Queensland in Australia, Papua New Guinea, Indonesian provinces. Naturalized in India.

Seeds of A. mangium with aril

Habit : A fast-growing, evergreen tree. Leaves (phyllodes) are large, dark green with commonly four primary longitudinal nerves (three in *A. auriculiformis*); leaves are compound in the juvenile stage.

Phenology : The time of flowering differs throughout its natural and planted range. The peak season is reported from July to August. Seed mature from January to April.

Uses: *A. mangium* is a major plantation species in Asia, grown mainly for the paper and pulp industry. Wood is also used as fuel wood; timber is used for furniture and construction purposes-trees are used in erosion control, shade, and shelter.

How to identify and Seed Handling: Fruit is an irregularly coiled, woody dehiscent pod. Seeds are black and shiny with a bright yellow-orange funicle attached to the pod. 66,000-120,000 seeds kg^{-1} (Awang and Taylor, 1993). Seeds are orthodox and viable for up to one year and more under natural conditions in a sealed tin. Seeds are soaked in boiling water for 30 seconds and followed by soaking in cold water for 24 hours for better germination. Germination percentage ranges between 75-90 percent with suitable treatments. The germination type is epigeal.

Acacia mearnsii De Willd.

Family : Fabaceae (Mimosoideae)

Synonyms : *A. molllissima* Willd., *A.decurrens* Willd.

Common name: Black wattle.

Distribution : It is native to Australia, Queensland, Tasmania, Victoria, but adapted to a wide range of sites from the temperate and subtropical lowlands to tropical highlands.

Seeds of A.mearnsii

Habit : A fast-growing, nitrogen-fixing tree or large shrub grows up to 6-10 m tall, sometimes reaching 25 m.

Phenology : Inflorescence is a simple or panicled raceme consisting of 5 to 10 pale yellowed heads. Flowering can be seen from February-April. Seeds mature during October- November.

Uses: High-quality tannins can be extracted from bark; wood is used in pulp, firewood, and charcoal production. Trees are grown to control soil erosion, windbreak, and soil improvement. It is a good source of green manure.

How to identify and Seed Handling: Fruit is a pod with finite hairs constricted between the seeds. Seeds are tiny, hard and black with short yellow- cream coloured aril. around 65,000-90,000 seeds weigh one kg. The seed is orthodox and stored for several years with proper storage methods. The seeds have a hard seed coat, requiring pre-treatment by immersion in boiling water for 1 to 2 min, followed by soaking in normal water for 24 hours. Germination is typically 50-80 percent (Anon., 1987).

***Adenanthera pavonina* L.**

Family : Fabaceae (Mimosoideae)

Common name: Coral tree, Peacock tree

Distribution : *Adenanthera pavonina* is found naturally in India, China, and Malaya Peninsula. In America, it has been introduced in Puerto Rico, Cuba, Jamaica, Trinidad, Venezuela, Brazil, Costa Rica, and southern Florida.

Seeds of A. pavonina

Habit : It is a medium to large sized deciduous tree, reaches up to 24m.

Phenology : Seeds mature during March- July.

Uses: *Adenanthera pavonina* has been planted as a shade and ornamental tree. The wood is solid and durable and used for firewood and construction. The seeds are used as beads in costume jewellery and bracelets. Raw seeds are poisonous.

How to identify and Seed Handling: Pods long and narrow, with slight constrictions between seeds, dark brown in colour and twist upon dehiscence. Seeds are showy, scarlet. Seeds adhere to the pod. There are 1500-3300 seeds per kg. Seeds are orthodox and stored for an extended period. Germination is epigeal and can achieve 72 percent germination by adopting suitable pre-sowing treatments like hot water or sulphuric acid scarification.

Adina cordifolia Roxb.

Seeds of Haldu

Family : Rubiaceae

Synonym : *Haldina cordifolia* (Roxb.) Risdd.

Common name: Haldu.

Distribution : Haldu is distributed in greater parts of India and Myanmar, Sri Lanka. It is commonly seen in the deciduous type of forest.

Habit : Haldu is a large deciduous tree growing up to 25 m.

Phenology : Flower heads can be seen from June to August. Fruiting is generally from October to January.

Uses: The wood is used for structural and turnery works and in traditional medicine to cure different diseases such as chronic cough and uses in jaundice, stomachache, fodder, and swelling in the stomach, the roots are astringent and constipating and are helpful in diarrhoea and dysentery.

How to identify and Seed Handling: Fruits are yellowish black with numerous small seeds. Seeds are oblong, brown in colour, minute, and shortly winged. About 90,00,000 to 1,00,00,000 seeds weigh one kilogram. Seeds are orthodox and can be stored for one year in ambient conditions. Germination is epigeal and up to 90 percent (Sen Gupta, 1937).

Aegle marmelos (L.) Correa

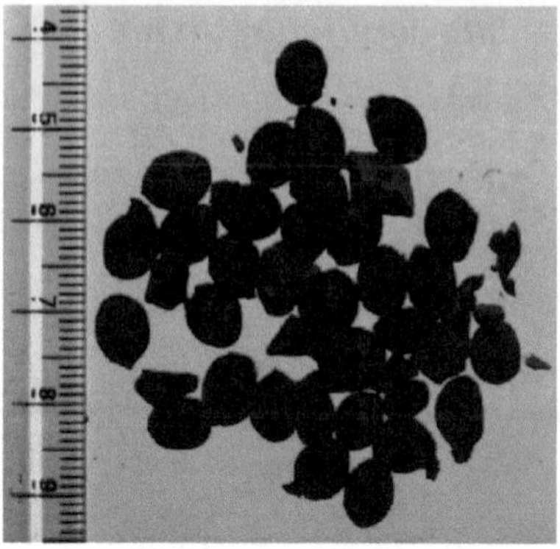

Seeds of Bael

Family : Rutaceae

Synonym : *Crataeva marmelos* L.

Common name: Golden apple, *Bael.*

Distribution : Bael is native to the Indian Subcontinent and Southeast Asia. It is distributed throughout India, except in Arid and Western Ghat's high rainfall zones.

Habit : Bael is a spinous deciduous and aromatic slow-growing tree with long solid and axillary spines.

Phenology : Flowering usually appears with young leaves. Seeds mature during March-May and October-November.

Uses: The fruit, leaves, bark is used in medicine. Leaves are used for sacred offerings among Hindu's. The fruits can be eaten either fresh or after being dried. It can be made into a beverage called *Bael pana.*

How to identify and Seed Handling: Fruit is a woody berry, slightly pear-shaped with a thick, hard rind and hard to break down. Seeds are oblong and compressed, with mucous testa embedded in the fruit pulp. About 4800 to 7200 seeds weigh one kilogram (Kindt and Salim, 1997). Seeds are recalcitrant and viable for a short period. Germination is hypogeal and 60 to 90 percent. Pre-treatment is not necessary for freshly collected seeds.

Alnus nepalensis D. Don

Seeds of *A. nepalensis*

Family : Betulaceae

Synonyms : *Alnus mairei* A. Léveillé, *Clethropsis nepalensis* Spach

Common name: Nepalese alder.

Distribution : Native to Pakistan, eastern Nepal, Bhutan, northern India, south-western China, upper Myanmar, and parts of Indochina

Habit: It is a large, fast-growing deciduous or semi-deciduous tree that reaches about 30 m in height and with a diameter of 60 cm.

Phenology: Flowers during September-October, and Fruits ripen during December- March.

Uses: Wood is an important source of firewood and charcoal. Trees are grown as shade trees in tea gardens. The bark is widely used for dyeing and tanning. The leaves are palatable to cattle and are mixed with other feeds.

How to identify and Seed Handling: The catkins resemble woody cones, held upright on short stalks. Seeds are light brown, circular, with membranous wings. There are typically 2.3-3.5 million seeds per kg. The seeds are orthodox. Viability can be retained for at least a year under hermit conditions. Seed emptiness is the primary cause of poor germination. Seeds do not need pre-treatment; germination reported is 70 percent (Carlowitz, 1991).

Alstonia scholaris R. Br.

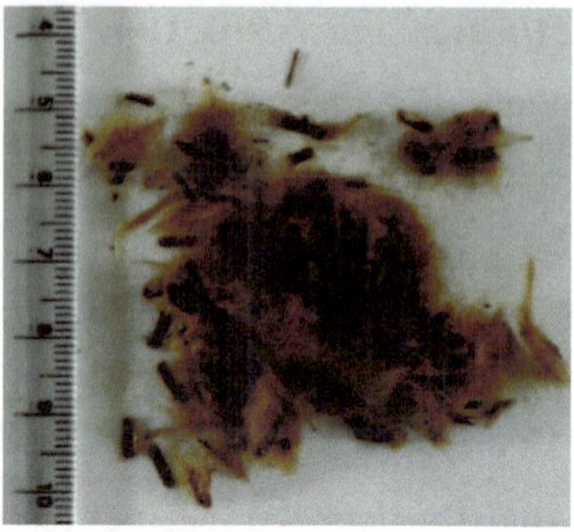

Seeds of A. scholaris

Family : Apocynaceae.

Synonym : *Echites scholaris* L.

Common name: Devils tree, Doddapale.

Distribution : Devils tree is commonly found in India's moist forests; in tropical semi-evergreen and moist deciduous forests. It is common in West Coast forests.

Habit : This is a moderately fast-growing evergreen tree, which can attain a height of 30 m with the fluted and buttressed stem.

Phenology : Flowering can occur from December to March, and seeds mature from May to July.

Uses: The wood is used in packing cases, tea boxes, and match stick making. It can also be used as plywood after necessary treatments. The bark is bitter and used as an astringent, tonic remedy for malaria, diarrhoea, and dysentery. The milky juice of bark and leaves is applied to sores and ulcers.

How to identify and Seed Handling: Fruits are paired oblong follicles. Follicles are slender and pendulous with numerous seeds. Seeds are linear-oblong shaped with a tuff of hairs at both ends. 2,70,000 to 6,60,000 seeds weigh a kilogram. Seeds are orthodox. Seeds germinate after a fortnight. Germination is epigeal. The germination percentage is 50 to 60 percent (Kumar and Bhanja, 1992). No treatment is required.

***Anacardium occidentale* L.**

Seeds of A. occidentale

Family : Anacardiaceae

Synonyms : *Acajuba occidentalis* Gaertn., *Cassuvium pomiferum* Lam.

Common name: Cashew.

Distribution : Native to north-eastern Brazil, cultivated in many tropical countries.

Habit : Anacardium is small to a medium-sized Evergreen tree that grows up to 6-12 m with crooked bole.

Phenology : Flowering varies from December to April from location to location. The nuts ripe from March to June.

Uses: Anacardium is mainly grown for its nut. The tree is suitable for agroforestry. The cashew apple has a high content of vitamin C. The oil extracted from the cashew nut shell (CNSL) is commercially valuable. The cashew apple is soft, juicy, and edible, giving a delicious beverage.

How to identify and Seed Handling: As the fruits mature, the flower stalk enlarges into the fleshy edible part, *i.e.*, the Cashew apple, becoming yellow to red at its maturity. The fruits consist of kidney-shaped nuts with two large cotyledons embedded in a hard shell. There are 240-300 nuts per kg. Seeds are orthodox and stored for more than a year (Shetty, 1972). There is no dormancy; fresh nuts can be sown directly without pre-treatment. That is recommended to soak the seeds overnight for better germination of stored seeds before sowing.

***Annona squamosa* Linn.**

Seeds of A. squamosa

Family : Annonaceae

Common name: Sugar apple,Custard apple.

Distribution : Native to Tropical America and West Indies. Most widely cultivated fruit throughout the tropics & warmer subtropics.

Habit : This is an evergreen, small to medium-sized tree grows faster. Start fruiting two to three years after planting.

Phenology : Flowering and fruiting throughout the year in its native range.

Uses: This is a multi-utility non-timber tree. It is mainly grown as a fruit tree, a medicinal and industrial product source. The tree is grown as an ornamental tree in tropical gardens. Root and leaves are used in traditional medicine.

How to identify and Seed Handling: Seeds are dark in colour, shining, and smooth with a hard seed coat. Seeds are short and durable or have no dormancy, which can lose longevity soon. Pre-treatment is necessary for germination. Scarification or water soaking is the best method to overcome the germination problem. Germination is slow, take one to three months to germinate (Adeniji, 2014).

Anogeisus latifolis Wall ex Bedd.

Seeds of axle wood

Family : Combretaceae

Common name: Axle wood, Dindal, Bakli

Distribution : Axle wood occurs throughout tropical and sub-tropical India except for northwest India, northeast India, and Andaman's. It is chiefly associated with *Terminalia* spp., *Mitragyna parviflora, Adina cordifolia, Diospyros* spp.

Habit : This is a medium to large-sized deciduous tree, attaining a height of 27 m and a diameter of 60 cm

Phenology : The tree remains leafless from about February to April. The small greenish-yellow flowers appear from June to September; fruits ripen from February to March.

Uses: Wood is heavy, strong, tough, and used in agricultural implements, tool handles, and cart axles. It is also used for making furniture, rafters, house constructions, etc. It also makes excellent charcoal. Leaves are used as fodder and also used to rear silkworms.

How to identify and Seed Handling: Fruit is a small drupe, orbicular, shining, compressed with a narrow wing on each side, one-seeded, yellowish-brown, and wedge-shaped. The tree seeds abundantly every year, but the fertility of the seed is very poor. There are over 1.00 lakh seeds per kg, but germination exceeds only 1 percent. The average germinative capacity varies from 0.05 percent to 4.62 percent, the average being 0.122 percent (Anon., 1950).

Seeds of Agarwood

Aquilaria agallocha Roxb.

Family : Thymelaeaceae

Synonym : *Aquilaria malaccensis* Lam.

Common name: Agarwood, Malayan eagle wood

Distribution : *A. malaccensis* occurs in India, Burma, Malaysia, the Philippines, and Indonesia. It is found in primary and secondary forests, mainly on lowland and hillsides.

Habit : Agarwood is an evergreen tree that grows up to 20-40 m tall.

Phenology : Trees flowering in July-August have mature fruits in November-December; trees flowering in March-April bear fruits in July-August.

Uses : Main use is agarwood, a highly appreciated and priced fragrant wood caused by aromatic resin accumulation. Wood without or low resin content can be used for boxes, interiors, or veneer. The inner fibrous bark has occasionally been used locally as raw material for clothing and ropes.

How to identify and Seed Handling: Fruits are egg-shaped capsules, leathery exocarps with fine hairs, having two seeds per fruit. Seeds are ovoid, blackish brown, with densely covered red-brown hair. About 1400 to 1800 seeds weigh one kilogram. Seeds are recalcitrant. Most seeds germinate within three weeks; the fresh seed should have about 70-80 percent ger-mination.

***Artocarpus hirsutus* Lam.**

Seeds of Hebbalasu

Family : Moraceae

Synonymous : *Artocarpus pubescens*, *Saccus hirsutus*

Common name: Aini, hebbalasu, Anjil

Distribution : These are endemic to the Western Ghats and fairly common in North Canara, Mysore, and Kodagu forests.

Habit : Large evergreen tree attaining a height of 35 m. The tree has white latex.

Phenology : Flowers from December to February. Fruits ripen in April-June.

Uses: Wood is moderately hard, durable, and yellowish, used in furniture and construction, boating, building, etc. Fruit and seeds are edible.

How to identify and Seed Handling: Fruits are yellow syncarp with long echinate. Seeds are ovate to oblong in shape, with white smooth coated. Seeds are recalcitrant, quickly lose viability, and are viable only for one month under natural conditions. About 1500 to 2300 seeds weigh one kilogram (Carlowitz, 1991). Germination is hypogeal and varies between 30 to 80 percent, based on the storage time without pretreatment (Sen Gupta, 1937).

***Azadirachta indica* A. Juss.**

Family : Meliaceae

Synonyms : *Antelaea azadirachta* (L.) Adelb., *Melia azadirachta* L.,

Common name: Neem, morgasa Indian lilac

Seeds of neem

Distribution : Native to India and Burma. Cultivated in most of the semi-arid and sub-humid areas of Asia, Africa, Australia, and the Southern United States.

Habit : A medium-sized Evergreen or Deciduous tree under drought conditions. Generally, it attains a height of 15 m, rarely up to 25 m.

Phenology : The tree remains in full foliage in summer. It flowers from January to March; fruits ripen from June to August.

Uses: The foliage is used as fodder. Traditionally, it is used for cheap furniture, farm implements, tool handles, etc. The oil extracted from seeds (Margosa oil) has medicinal properties. They are used in soap making-neem cake is used as an insecticide. Gum and fibre are extracted.

How to identify and Seed Handling: The fruits are drupes, green/yellow when ripe, has juicy fruit pulp. The pyrene contains one or two seeds. About 8000-9000 pyrene per kg. Neem seeds are recalcitrant and have short viability. Germination is epigeal and up to 90 percent (Sen Gupta, 1937). Pre-treatment is not required for fresh seeds. Depulping and cleaning the seed considerably improves the germination percent.

Bauhinia purpurea Lam.

Seeds of B. purpurea

Family : Leguminosae

Subfamily : Caesalpinioideae

Common name: Geranium tree.

Distribution : It is widely distributed in India. It occurs in the Himalayan tract to southwards in tropical wet- evergreen, semi-evergreen and moist deciduous forest types.

Habit : This is a moderately growing, medium-sized evergreen tree.

Phenology : Flowers appear from September to December; pods are ripe from January to May.

Uses: *B. purpurea* is an excellent fodder tree. Root, bark, flowers, and gum have medicinal value. Seeds contain non-drying oil. Wood is moderately strong and rigid, used as timber, packing cases, agricultural implements, and fuel wood. It is an ornamental tree. Flowers are cooked as vegetables and pickles.

How to identify and Seed Handling: Fruit is a pod consisting of 12 to 16 seeds that burst after ripening. Seeds are oblong, smooth, brownish, and compressed. Around 6500-7500 seeds weigh a kilogram (Ghosh, 1977). Seeds are orthodox and stored up to 8-12 months in airtight containers. Germination capacity is 85-100 percent. Fresh seeds perform better without any treatment. Stored seeds are soaked in hot water for better germination.

***Bauhinia racemosa* Lam.**

Seeds of B. racemose

Family : Leguminosae

Synonym : Piliostigma racemosa (lam.) benth

Subfamily : Caesalpiniodeae

Common name: Kanchan, Mandara

Distribution : It is common thoughtout India. It occurs in dry and moist deciduous forests. It is also cultivated in Uttar Pradesh, Madhya Pradesh, Orissa, Gujrat, West Bengal, and Punjab.

Habit : Mandara is a slow-growing, small to moderate-sized deciduous tree with a spreading crown.

Phenology : Flowering can be seen from March to June, and pods will start ripening in November to December.

Uses: Wood is a good source of fuel. It is also used locally for agricultural implements. Bark yields dye and fibre. The gum extracted is used in traditional medicine. The seeds are roasted and eaten.

How to identify and Seed Handling: Fruit is a pod consisting of 12-20 seeds. Seeds are collected by beating the sundried rigid pods with a wooden mallet. Seed weight varies from 7000-8000 seeds per kg. Seeds are orthodox, and they can be stored for about one year. Seed testa is hard; hence boiling water treatment is advised for better germination. Germination capacity is 25-95 percent. Fresh seeds give better germination than stored seeds. Germination is epigeal.

Butea monosperma (Lamk) Taub.

Seeds of Muttuga

Family : Leguminosae-Faboideae

Synonym : *Butea frondosa* Roxd.

Erythrina monosperma Lam.

Common name: Palas, Muttuga, Flame of the forest.

Distribution : Palas are found in most parts of India. It is common in mixed dry deciduous forests of central and Southern India. It is typical of open grassland and scrubs jungle.

Habit : Moderately fast-growing, medium-sized deciduous tree with a crooked trunk.

Phenology : Flowers are orange-scarlet colored and appear from February to April. Trees are entirely leafless during flowering. Pods ripen in May-June.

Uses: Used as a host tree for lac cultivation. Wood is a good fuel source used as gun powder charcoal- a source of fodder. Powdered seeds are used as insecticides and medicine. Bark yields fibre. Bengal kino or Butea gum is extracted. Eating plates are prepared from leaves.

How to identify and Seed Handling: Fruit is a pod; broadly linear, tomentose contain flat, reniform, single seed at the apex. Around 1000-1200 seeds weigh one kg. Seeds are orthodox and stored for about a year. Germination is 70 percent to 90 percent (Sen Gupta, 1937). Germination is hypogeal. De-winged seeds germinate quickly without any pre-treatment.

***Calamus rotang* L.**

Family : Arecaceae

Common name: Rattan, Cane.

Distribution : *C. rotang* is restricted to plains and coastal areas of South Indian states, i.e., Karnataka, Kerala, Tamil Nadu, and Andrapradesh.

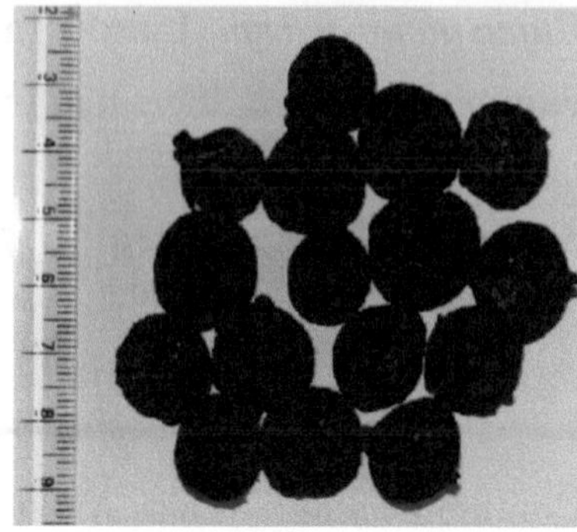

Seeds of C. rotang

Habit : This is a clustering; slender rattan stems are 1 cm in diameter without sheaths.

Phenology : ***Flowering:*** September- October.

Fruiting: March-May.

Uses: It is reported to be an excellent quality cane but not available in sufficient quantities. They are used in the furniture industry. Fruits are edible.

How to identify and Seed Handling: Fruit is fleshy, yellow in colour, and ovoid with scales. Fruits are small, and about 3500 to 4000 seeds weigh one kilogram. Seeds are primarily recalcitrant and lose viability within one month. Seeds can be stored in moist sawdust for up to one month. Germination is an adjacent ligular type. Germination percentage varies from 70 to 90 percent.

Calliandra calothyrsus

Seeds of calliandra

Family : Leguminosae,

Subfamily : Mimosoideae

Common name: Calliandra.

Distribution : It is native to Central America, Java, & Indonesia. It has been planted as an ornamental plant.

Habit : Calliandra is a fast-growing shrub to a small-sized tree attaining about 10 m in height.

Phenology : Flowering remains more or less throughout the year.

Uses: Leaves are used as fodder to feed the cattle. The charcoal has a high calorific value; fuelwood is used for domestic use. Wood is fit for pulp and paper manufacture. The tree is grown as an ornamental plant. Due to its flowering habit, it has the potential for apiculture.

How to identify and Seed Handling: Fruit is a pod, which dehisce on drying. Each pod contains 5 to 8 seeds. The seeds are treated with hot water and then soaked in cold water for 24 hours for better germination. Germination percentage is 70 to 85 percent.

***Calophyllum inophyllum* L.**

Seeds of C.inophyllum

Family : Clusiaceae (Guttiferae)

Synonyms : *Balsamaria inophyllum* (L.) Lour.

Common name: Polang, Pinnai, Alexandrian laurel, Borneo mahogany

Distribution : It is indigenous to India, Malaysia, Indonesia, and the Philippines. Commonly found in evergreen and semi-evergreen forests up to an altitude of about 1550 m.

Habit : It is a medium to a large-sized evergreen tree that reaches up to 15-25 m with a cylindrical bole.

Phenology : Flowering begins from March to June; fruits ripen from September to December.

Uses The decorative timber produced for cabinetwork, parquet flooring, construction purposes, musical instruments, veneers, plywood, canoes, and boats. It is a famous ornamental tree grown in the urban street. The oil extracted (tamanu oil, pinnay oil, or domba oil) from the seed is used in medicines.

How to identify and Seed Handling: The fruit is an ovoid drupe. Seeds are large, surrounded by endocarp. There are 450-500 seeds per kg. It has short viability, but it is uncertain due to the high oil content or because the seeds are recalcitrant. Usually, the seed can be stored for up to one year; after that, viability will drop. Germination is hypogeal and more than 90 percent. For quick and better germination seed coat is removed before sowing.

Canarium strictum Roxb.

Seeds of C.strictum

Family : Burseraceae

Common name: Black dhup, Raal dhup, Black dammar.

Distribution : It is endemic and indigenous to India's Eastern and the Western Ghats. This is spread over South and South East Asia.

Habit : Black dammar is a slow-growing, large, resinous species.

Phenology : Flowering will be seen in March and April, and the fruit set can be notable in late April and onwards.

Uses: This is commercially harvested for the dammar (resin). It is used traditional medicine, and it has spiritual importance.

How to identify and Seed Handling: Seeds are nuts with flesh epicarp and stony endocarp. They are orthodox, having long viability. Seed weight ranged from 2.6 to 6.5 grams. Cow dung slurry-treated seeds germinated better (52 %) than other treatments (Scarification-20 %, GA3 -5.6 %) (Kala *et al.,* 2014). Seeds passed through animals (deer members) and birds (Indian grey hornbill and pied hornbill are primary dispersers) digestive system, readily regenerate, and the germination percentage is also highest.

***Careya arborea* Roxb.**

Seeds of kaval tree

Family : Lecythidaceae

Synonyms : *Barringtonia arborea* (Roxb.) F. Muell.

Common name: Wild Guava, Kaval.

Distribution : Tree found sporadically throughout the greater part of India and Burma.

Habit : *Careya arborea* is a small to medium-sized deciduous tree growing up to 15 m tall.

Phenology : Flower blooms from March to April. Fruit ripe during June

Uses: The plant is a popular herbal remedy in India, where it is often harvested from the wild for local use and trade. Flowers and young leaves are eaten as green vegetables. The tree also supplies dye, fibre, gum, tannins, and kindling for limited practice. It is sometimes grown as a shade avenue and ornamental.

How to identify and Seed Handling: Fruit is a fleshy berry containing several seeds. Seeds are exalbuminous. Seeds are round or oval, brown. Seeds are intermediate and cannot be stored for a more extended period. About 2000 to 4000 seeds weigh one kilogram. Germination is up to 90 percent, and germination is complete in 11 to 46 days. Soaking seeds in water hasten the germination.

***Cassia fistula* Linn.**

Seeds of C. fistula

Family : Leguminosae

Subfamily : Caesalpinioideae

Synonym : *Cassia rhombifolia* Roxb.

Common name: Indian laburnum, Konne, Raj brikh.

Distribution : It is most widely distributed in India's tropical moist and dry deciduous forests. It is also found in Myanmar and Sri Lanka.

Habit : Kakke is a slow-growing medium-sized deciduous tree producing a drooping yellow flower.

Phenology : Flowering can be seen chiefly with the emergence of new leaves from April to July. Pods ripe during December continue till March and start falling during April to May.

Uses: Timber is used in agricultural implements and posts. It is grown as an ornamental tree in avenues and parks. Leaves and flowers have medicinal value; flowers are used in traditional culture. Edible gum can be extracted from the sticky pulp of the matured pod.

How to identify and Seed Handling: Fruit is a woody pod. Seeds are embedded in a soft and sticky pulp. Seed weight varies from 6000- 7000 per kilogram. Seeds are orthodox. The hard and impermeable testa avoids loss of viability. Seeds are orthodox and can be stored for up to 10 to 14 years (Dent, 1948). Germination percentage is up to 65 percent. Epigeal type of germination is observed. Because of hard testa, need pre-treatment to improve the germination. Soaking of seeds in boiling water for 5 minutes and then in cold water for 24 hours, or scarification with Conc. Sulphuric acid is advisable for better germination.

***Casuarina equisetifolia* L.**

Seeds of C. equisetifolia

Family : Casuarinaceae

Synonyms : *Casuarina littoralis* Salisb., *C. littorea* Oken., *C.muricata* Roxb.

Common name: Beefwood, Coast She-oak.

Distribution : The species naturally occurs in subtropical and tropical coastlines from Northern Australia throughout Southern Myanmar and Thailand. It has been introduced worldwide in the tropical and subtropical zone.

Habit : Beefwood is an evergreen Nitrogen-fixing, fast-growing tree having needle-like drooping branchlets. Leaves are tiny and reduced, arranged in whorls.

Phenology : Flowering and fruiting will occur once or twice in the year based on location in between December and June (Rai, 1999).

Uses: Casuarina is a multipurpose tree species. It is popularly known for "the best firewood in the world" and produces high-quality charcoal. Species are tolerant to salt, which is used for erosion control in coastal areas. It is also used in paper pulp, timber, land reclamation, and soil improvement.

How to identify and Seed Handling: Fruits are woody cones. Seeds are grey or yellow-brown winged samaras or achene. There are 350,000- 750,000 seeds per kg. Seeds are viable for up to a year with hermetic storage conditions. No pre-treatment is required. Germination is epigeal, and the germination percentage is 50 percent (Subramanian *et al.,* 1992).

Ceiba pentandra (Linn.) Gaertn.

Seeds of kapok

Family : Malvaceae

Synonyms : *Bombax pentandrum* Roxb., *Eriodendron anafractuosum* DC.

Common name: Kapok, cotton silk tree.

Distribution : Naturally occurring in the moist evergreen and deciduous forests of the United States, Central America, and South America. They are naturalized throughout India.

Habit : Kapok is an elegant deciduous tree with a clear bole up to 12-15 m, with a buttress having woody conical prickles when young.

Phenology : Flowering occurs between December and February when the new leaves appear. Fruiting occurs till the end of June.

Uses: The flosses are used as excellent cushion material for many purposes; the foliage is used for fodder for livestock, and kapok oil extracted from the seeds is used for industrial purposes. Wood is used for boxes, crates, plywood, pulp, and paper products.

How to identify and Seed Handling: Fruits are leathery capsules. Capsules split open into five valves, revealing a mass of woolly hairs in which the 120-175 seeds are embedded. Seeds are black to dark brown, having 20-25 percent oil. There are 10,000-45,000 seeds kg^{-1} (Gupta, 1993). The seeds are orthodox. They contain large amounts of oil that tend to go rancid quickly, and the viability diminishes rapidly. No pre-treatment is required; fresh seeds usually germinate 90-95 percent.

Dalbergia latifolia Roxb.

Seeds of Rosewood

Family : Leguminosae

Subfamily : Faboideae

Synonym : *Dalbergia emarginata* Roxb.

Common name: Rosewood, Beeter, Shisham.

Distribution : Occurs in the Sub-Himalayan tract to Central, Western, and Southern India in a moist deciduous forest. In South India, it is found in semi-evergreen and deciduous forests.

Habit : Rosewood is a slow-growing, large deciduous tree, attaining a height of 25 m.

Phenology : Flowers appear from February to April. Mature fruits are available from December to May.

Uses: The wood can be peeled or sliced to make decorative veneer and plywood. It is fit for high-grade plywood and, owing to its attractive color and figure, for decorative veneer. The wood is used for fixtures, cabinet making, and decorative. It is suitable for turnery and is excellent for high-class bentwood furniture, walking sticks, umbrella handles, and other articles.

How to identify and Seed Handling: Fruit contains 1 to 3 compressed, deep brown to black seeds. About 21000 to 40000 seeds weigh one kilogram. Seeds are orthodox; seeds remain viable for up to 6 months. Seeds have no dormancy, so pre-treatment is unnecessary, although pre-soaking in water for 12-24 hours can accelerate germination. The germination rate is 45 to 80 percent.

Dalbergia paniculata Roxb.

Pods of D. paniculata

Family : Leguminosae

Subfamily : Faboideae

Synonyms : *Dalbergia lanceolaria* L.f.

Common name: Takoli, Hasirugani

Distribution : It is widely distributed in the dry and moist Deciduous Forest throughout the Sothern and Central India.

Habit : Moderately fast-growing, medium-sized deciduous tree with a crooked truck, attaining a height of 15 to 20 m.

Phenology : Flowers are minute and appear from February to April. Matured fruits are available from September to October.

Uses: Timber is locally used in agricultural tools. Foliage is lopped as fodder to feed cattle. Wood is also used as timber and fuelwood. The tree is grown as greenbelts and in reforestation sites.

How to identify and Seed Handling: Fruit is a pod, flat with narrow ends. Seeds are brown with a smooth texture. Around 7000 to 8000 seeds weigh one kilogram. Seeds are intermediate and viable for up to six months. Germination is epigeal and up to 90 *Dalbergia paniculata.* Germination starts after seven days and continues till 25 days.

Dalbergia sissoo Roxb. ex DC.

Pods of Shisham

Family : Leguminosae

Synonyms : *Dalbergia pendula, Pterocarpus sisso*

Common name: Shisham, Indian Rosewood

Distribution : It is not a tree of Deccan; it occurs throughout the Sub-Himalayan tract and Assam valleys. Trees are also found in Punjab, Karnataka, Uttar Pradesh, Tamil Nadu, etc.

Habit : Shisham is a medium to large-sized deciduous tree, attaining a height of up to 30 m.

Phenology : Young flower buds appear along with the new leaves during March and April. Pale green pods form rapidly by the end of April and ripen from November to December.

Uses: The wood of shisham is used for permanent structures, primarily for door and wind shutters, wooden flooring, and paneling. It is also preferred in building railway coaches, trucks, and lorry bodies. Leaves are used as fodder and also act as a stimulant.

How to identify and Seed Handling: Pods are light, flat, and indehiscent, which remain dormant during hot months. About 52000 to 55000 seeds weigh one kg. Germination capacity is 85 to 90 percent (Anon., 1962). The germination is epigeous. Pre-treatment is unnecessary; however, breaking the pods into segments and soaking them in cold water improves germination.

Delonix regia (Bojer. ex Hook.) Rafin.

Seeds of Gulmohar

Family : Fabaceae,

Synonym : *Poinciana regia* Bojer. ex Hook.

Subfamily : Caesalpinioideae

Common name: Glod Mohar, Gulmohar.

Distribution : It is native to Madagascar. It is planted as an avenue tree in dry and moist regions of tropical India.

Habit : Gulmohar is a fast-growing broad-crowned deciduous, ornamental tree.

Phenology : Flowering will start in April and continue until the rain break. Pods will remain till the following year of flowering.

Uses: Gulmohar is extensively cultivated in avenues and gardens for shade and ornamental purposes. Seeds contain gum that may find use in the textile and food industries (CSIR, 1986).

How to identify and Seed Handling: Fruit is a long, flat, woody pod. Seeds are oblong, brownish, and marked with spots transversely. Around 2500-3000 seeds weigh one kilogram. Seeds are orthodox and can be stored for up to one year. Germination is epigeal. Germination percentage varies from 25-68 percent. Seeds have hard testa and take time to germinate. Stored seeds will perform well when they are treated by soaking in boiling water.

Dendrocalamus strictus **(Roxb.) Nees**

Seeds of D. strictus

Family : Poaceae (Graminae)

Synonym : *Bambus stricta* Roxb.

Common name: Male bamboo, solid bamboo, Kiribidiru.

Distribution : The most distributed bamboo of India, throughout the deciduous forests.

Habit : Densely tufted, unarmed deciduous bamboo.

Phenology : The species flowers gregariously, followed by mass mortality occurring after 30 years. Flowers are produced from November to March. Seeds ripe in March-April.

Uses: The most used bamboo species for construction, making baskets, mats, furniture, flooring, tool handles, ladders, etc. It is an important raw material for the paper and rayon industries. Leaves are used to feed cattle and elephants. Young shoots are used in curries and pickles.

How to identify and Seed Handling: Caryopsis is brown and has an aristate rostrum at the apex. Seeds weight varies from 25,000-30,000 seeds per kilogram. Seeds are orthodox and viable for up to one year if stored in sealed tins. The germination capacity of fresh seeds is about 75 percent. Germination is epigeal. Cold water treatment; soaking seeds for 24 hours is recommended for better germination.

Dipterocarpus indicus Bedd.

Seeds of Gurjan

Family : Dipterocarpaceae

Synonyms : *Dipterocarpus turbinatus* auct.

Common name: Gurjan, Kalpine.

Distribution : Gurjan is an evergreen tree found in Southern ghats from North Canara southwards, all along the eastern ghats to Kerala and Tamil Nadu. It is common in Karnataka and Kerala.

Habit : Gurjan is a large evergreen tree found in evergreen forests that can attain a height of 47 m with a clean, cylindrical bole.

Phenology : Flowers will appear from December to January. Fruit ripens from April to June.

Uses: Timber is widely used in plywood industries and used in scantlings, planking, ship building, masts, and heavy packing cases. It is good fuel wood. Treated timber is used for railway sleepers.

How to identify and Seed Handling: Fruit is a winged samara. Fruits are heavy with large woody samaras. About 170 fruits weigh a kilogram. The viability is very low and can be stored for a month. Since fruits are produced in the rainy season, natural regeneration fails. Germination is hypogeous. The germination capacity of freshly collected seed is reported at 90 percent. Although, based on seed and storage status, the germination varies between 40-90 percent (Sen Gupta, 1939).

Ensete superbum (Roxb.) Cheesman

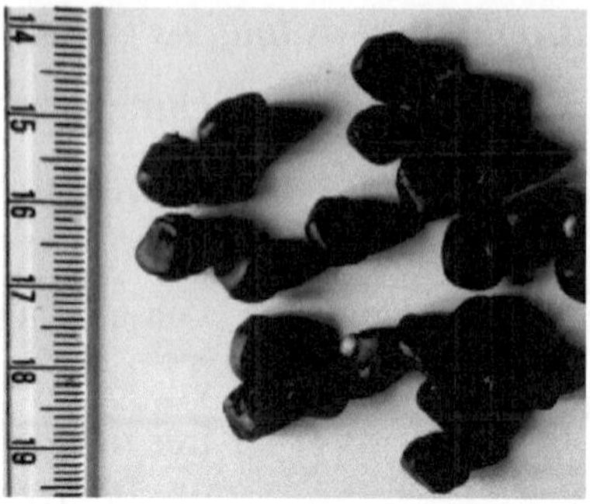

Seeds of rock banana

Family : Musaceae

Synonyms : *Musa superba* Roxb

Common name: Rock Banana, Wild Plantain

Distribution : This is a wild banana species found only in India (and probably in Thailand). It's very hardy a prefers rocky, barren areas.

Habit : It is characterized by 3 to 5 m high, pseudo-stem, waxy, and does not produce suckers.

Phenology : Flowering and fruiting from June to September.

Uses: Core inflorescence and the pseudo stem is eaten as a salad while the flower is cooked as a vegetable. Seeds are medicinal properties used to treat kidney stones and diabetes.

How to identify and Seed Handling: Single plant can produce 60 to 100 fruits of variable sizes- about 6 to 45 seeds embedded in a single fleshy fruit. The seeds are black, angled with predominant depression, light in weight, and have a thick and tough seed coat. The seeds show dormancy and do not germinate during dry seasons. Germination can improve up to 64 percent by treating the seeds with hot water.

Enterolobium cyclocarpum (Jacq.) Griseb.

Seeds of Enterolobium

Family : Fabaceae - Mimosoideae

Synonyms : *Albizia longipes* Britton & Killip,

Pithecellobium cy-clocarpum Mart.

Common name: Ear pod tree, elephant ear, monkey-soap.

Distribution : This is naturally distributed in central Mexico, Central America, Brazil's. It occurs in a wide range of tropical forest types.

Habit : This is a large, deciduous tree that grows up to 40 m tall

Phenology : Within its native range, flowering occurs in March-April, and fruits fall in April-May.

Uses: The species is mainly used as a pasture tree in silvo-pastoral systems. The pods are used for fodder; they are highly palatable. It is used in house building and for indoor elements such as paneling.

How to identify and Seed Handling: The fruit is an indehiscent pod, which resembles an ear, with 10 to 15 seeds embedded in sweet pulp. Seeds are 15 to 20 mm long, flat, dark, or reddish-brown. There are around 900-1400 seeds per kg. The seeds are orthodox and stored at 10 percent moisture content in hermetically sealed containers, which will be viable for several years. Germination is good, about 85 percent. The seed coat is extremely thick; seeds will not germinate unless scarified to allow water to enter the seed.

***Erythrina indica* Lam.**

Family : Leguminosae

Subfamily : Faboideae

Synonym : *Erythrina varigata* Roxb.

Common name: Indian coral tree.

Distribution : Is a species of Erythrina is native to the tropical and subtropical regions of eastern Africa, the Indian subcontinent, and northern Australia.

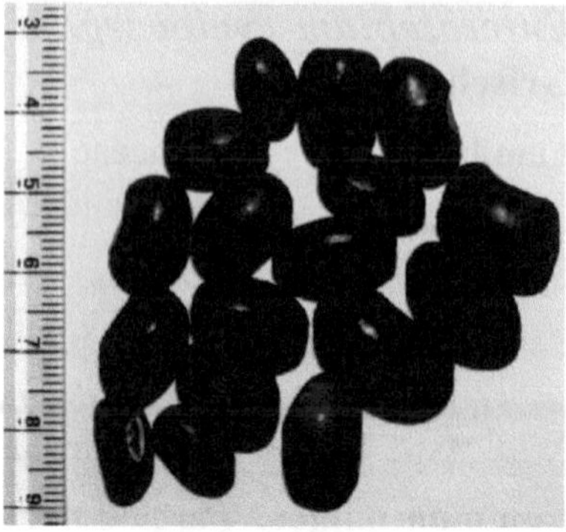

Seeds of E. indica

Habit : This is a fast-growing, deciduous tree with small black prickles and smooth greenish bark.

Phenology : The flowers appear from April to May, and the pods ripen during June.

Uses: The leaves and bark are used as cures in many South-East Asian countries. The young, tender leaves and sprouts are eaten as vegetables. A decoction of bark and leaves is used to treat dysentery. The wood has been tested as a source of pulp for the paper industry. Wood is used for making statues, toys, etc.

How to identify and Seed Handling: Fruits are pods. Seeds are bean-like, shining, and brown. Seeds are orthodox and stored for more than one year. Germination rates are generally high and are often 100 percent. Soaking seeds in hot water or abrading seed coat fasten the germination. Seeds germinate within 8 to 10 days of sowing.

***Eucalyptus tereticornis* Sm.**

Seeds of red gum

Family : Myrtaceae

Synonym : *E. umbellate* auct. non (Gaertn.) Domin

E. subulata A. Cunn. Ex Schauer

Common name: Mysore gum.

Distribution : It is an exotic from Australia. It has been hybridized and naturalized in most tropical countries.

Habit : Eucalyptus is a fast-growing evergreen tree with a straight bole and grows up to 45 m in height.

Phenology : Flowering occurs twice a year, first in May-June and again in October-November. The capsules are usually ripped within 1-2 months of flowering.

Uses: Wood is suitable for small timber, poles, packing casers, beams, and posts. It is being used even for cheap furniture. Wood is used for making pulp, wrapping, writing, and printing. The bark is suitable for the production of oxalic acid. Leaves have essential oil, which contains Cineole-medicinal eucalyptus oil.

How to identify and Seed Handling: Fruits are capsules. Capsules are hemispherical and disc wide. Seeds are extremely minute; about 3,67,400 seeds weigh a kilogram (Luna, 1996). Seeds are orthodox and are viable for about two years in sealed tins. Germination is up to 90 percent and is epigeal type.

Seeds of Gliricidia

Gliricidia maculate **var.** ***multijuga*** **Micheli**

Family : Fabaceae

Subfamily : Faboideae

Synonyms : *Gliricidia sepium* (Jacq.)

Lonchocarpus roseus (Miller) DC.,

Robinia sepium Jacq.

Common name: Mexican lilac, gliricidia

Distribution : It is native to Mexico and Central America but has now been introduced and naturalized all over the tropics.

Habit : Gliricidia is a small, fast-growing, and deciduous tree. It is a single to multi-stemmed tree with an open crown.

Phenology : Flowering and fruiting occur when the tree is leafless in the dry season.

Uses: Gliricidia is a multipurpose tree commonly planted as a live fence. Fodder is used to feed cattle- excellent green manure, rich in nitrogen. Wood is used in small agricultural implements. Seeds are excellent rat poison. It is called Mother of Cocoa since used as a shade tree for cocoa, coffee, and tea. Wood has a calorific value of 4900 kcal kg^{-1}.

How to identify and Seed Handling: Fruits are pods. Seeds are ovoid and brownish. The number of seeds per kilogram differs between 9500-10000 seeds. Seeds are orthodox and can be stored for about a year. The seeds have no dormancy, and hence, pre-treatment is not necessary. The germination percentage is 90 percent. For stored seeds, it is advised to soak seeds in hot water before sowing.

Gmelina arborea Roxb.

Seeds of G. arborea

Family : Verbenaceae.

Common name: White teak (Eng.), Gamari, Shivane

Distribution : Tree is native to Nepal, India, Sri Lanka, Bangladesh, Myanmar, Pakistan, Thailand, Vietnam, and Southern China. This is usually found in evergreen forests of Myanmar and Bangladesh and deciduous forest types of India.

Habit : It is a medium-sized fast-growing deciduous tree.

Phenology : In natural distribution areas, flowering starts in the dry season, and mature fruits will be available after 1½ months of flowering, usually from April to July.

Uses: The wood is used for light construction, planking, boxes, and carpentry-leaves are used as fodder for cattle.

How to identify and Seed Handling: The matured fruit (drupe) will turn yellow color or fall before turning yellow. Each fruit will contain 2-3 stones. About 1200-1500 seeds weigh one kilogram. Seeds can be stored for up to one year by maintaining 50 percent of viability. Fresh stones will show better germination without any pre-treatment. Seeds stored for an extended period can be soaked in water for 48-72 hours or imbibed in the dark, moist sand for 20 days to hasten germination. However, better germination can also achieve by weathering the seed. Germination is epigeal and varies from as low as 13 percent to as high as 90 percent (Troup, 1921).

Holoptelea integrifolia Planch

Fruits of Tapsi

Family : Ulmaceae

Synonym : *Ulmus integrifolia* Roxb.

Common name: Indian Elm, Kanju.

Distribution : This is throughout India from the sub-Himalayan tract to ascending 600 m elevation. Commonly found in the mixed deciduous forest of Peninsular India.

Habit : This is a large deciduous tree, which attains 25 m high with a spreading crown.

Phenology : Flowers appear during the leafless period from February to April, and fruits ripen soon after April and May.

Uses: The wood is used for indoor building purposes, low-cost furniture, cabinet work, packing cases, match boxes, bobbins, and agricultural implements. It is also used for plywood and paper-pulp. Leaves lopped for fodder to feed cattle.

How to identify and Seed Handling: Fruit is an oval or subordinate samara with a membranous wing, single-seeded, and seed is flat. About 25000 to 28000 seeds weigh one kg. Fresh seeds have a high germination percentage above 60 percent (Ghosh, 1977), deteriorating quickly after about months. Germination is epigeous. No treatment is required for freshly collected seeds.

Hopea parviflora Bedd.

Seeds of H. parviflora

Family : Dipterocarpaceae

Common name: Iron wood of Malabar, Kiralbogi.

Distribution : It grows in Western Ghats' tropical rain forests from Karnataka, Kerala, and Tamil Nadu.

Habit : *Hopea parviflora* is a slow-growing, gregarious evergreen tree. It can reach a height of 30-40 m. It is a shade demander in the seedling stage and requires full sunlight in later stages.

Phenology : Flowering can be seen from January to February. Matured fruits will be available from May to June.

Uses: Wood is hard, heavy, and highly durable. The timber is used in construction, ship and boat building, agricultural implements, ladders, decorative, etc. Its bark is a good tanning material.

How to identify and Seed Handling: Nut-like fruit with two wings will turn brownish from green color upon maturity. About 3000 to 3500 seeds with wings weigh one kilogram. Seeds are recalcitrant and viable for up to ten days under normal conditions. Germination is epigeal. Germination is 65-85 percent (Tompsett and Kemp, 1996).

Jacaranda mimosifolia **D. Don**

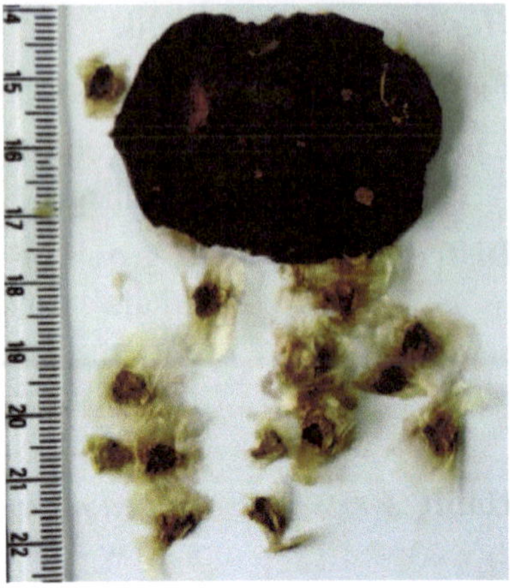

Seeds of Jacaranda

Family : Bignoniaceae

Synonym : *Jacaranda ovalifolia* R.Br.

Commonname : Brizilian rosewood, acaranda

Distribution : It is native to South America. In India, it was introduced as an ornamental avenue tree.

Habit : Jacaranda is a small to medium-sized evergreen tree attaining a height of 15 m or more.

Phenology : Tree flowers in the month of March-April. The matured seeds will be available from November to January.

Uses: Tree is mainly grown in parks and avenue roads as an ornamental tree for its handsome foliage and attractive flowers. The wood is locally used for tool-handles.

How to identify and Seed Handling: Fruit is a woody capsule. Capsules are broad and ovate with many compressed winged seeds. Around 2,50,000-2,80,000 seeds weigh one kilogram. Seeds are orthodox, and viability can be retained for one year. Germination is epigeal. Germination is 50 to 90 percent (Rai *et al.*, 1986). Soaking seeds in water for 24 hours will fasten the germination.

***Jatropha curcas* L.**

Seeds of Jatropha

Family : Euphorbiaceae

Synonym : *Curcas purgans* Medic.

Common name: Physic nut, purging nut.

Distribution : It is believed to be native to Mexico and Central America. It has been introduced to Africa and Asia and is now culti-vated worldwide.

Habit : Mall tree or large deciduous shrub, growing up to 8 m tall.

Phenology : Jatropha will flower for two seasons: April to June and July to November.

Uses: The species is widely grown as living fences. The seed oil is used as a biodiesel substitute for diesel oil. Press cake made from the plant is used as organic manure. The seeds contain 30-35 percent oil used as an insecticide for soap production and numerous other purposes.

How to identify and Seed Handling: Fruits are grey-brown capsules, customarily divided into three cells, each containing one seed. Seeds are black and oblong. There are 1700 to 2400 seeds per kilogram. Seeds are orthodox; the seeds cannot be expected to store for as long as most orthodox species because of high oil content. Dry seeds will normally germinate readily without pre-treatment. Germination is fast, and under suitable conditions, it is com-plete in 10 days. Germination is epigeal.

Khaya senegalensis (Desr.) A. Juss.

Seeds of African mahogany

Family : Meliaceae

Synonym : *Swietenia senegalensis* Desr.

Common name: African mahogany

Distribution : The natural distribution area is from Senegal to Su-dan and Uganda, south to the rainforest zone. It oc-curs in riverine forests and is scattered in high-rainfall savannah woodland. Plantations have been established in Asia and Latin America.

Habit : A deciduous tree, 15-20 m tall, reaching 35 m on fertile soils.

Phenology : Flowering occurs from November to December; fruits mature from February to May.

Uses: the wood is used for carpentry, joinery, fixture, cabinet work, ship construction, and decorative veneer. It is suitable for structure, carpeting, interior, vehicle bodies, toys, *etc.* It is also used as fuelwood and charcoal. The bitter-tasting bark is highly used in traditional medicine.

How to identify and Seed Handling: Seeds are brown, broadly ellipsoid to flat in shape, thin, and with narrowly winged margins. Seeds are orthodox. It is advised to add ash during storage to reduce insect pest attacks. About 2500 to 7000 seeds weigh one kilogram. Germination is up to 90 percent, and freshly collected seeds have better germination and are complete after two weeks. Germination is epigeal. Soaking stored seeds in water for 24 hours improve germination.

Lagerstroemia microcarpa Wight

Family : Lythraceae

Synonyms : *Lagerstroemia thomsonii* Koehne

L. lanceolata Wall. *ex* Cl.

Common name: Benteak, Bilinandi.

Seeds of Benteak

Distribution : Benteak is widely distributed on the west coast of the Indian peninsula. It comes in moist and dry deciduous forests, often mixed with teak, sal, and other forest crops.

Habit : Medium to large-sized deciduous tree, attaining a height up to 30 m.

Phenology : Flowering occurs from April to June, and matured capsules will be available from December to February.

Uses: Wood is an excellent timber source for construction, furniture, ship, and boat construction. Leaves are used as fodder.

How to identify and Seed Handling: Fruit is a capsule, dehisce after maturity, each capsule filled with winged cultriform seeds. Seed weight varies between 256000 to 280000 seeds per kg. Seeds can be stored for up to six months. Seed emptiness is high. Germination percentage is very less, i.e., 2- 20 percent without any treatment. Germination is epigeal.

Lagerstroemia speciosa **(Linn) Pers.**

Seeds of Hole dasavala

Family : Lythraceae

Synonym : Lagerstroemia reginae Roxb.

L. flos-reginae auct. non Retz.

Common name: Queen crape myrtle, Hole dasavala.

Distribution : occurs widely in India, Myanmar, Sri Lanka, and Bangladesh. It is found in semi-evergreen and moist deciduous forest types.

Habit : This medium-sized deciduous tree is well known for its beautiful flowers.

Phenology : The matured fruits are generally available from December to February, but capsules will be present in the tree for longer.

Uses: This is mainly planted as an ornamental tree. The wood is used in making scantlings, beams, posts, etc., and for country-made guns and rifles.

How to identify and Seed Handling: Fruit is a woody capsule filled with angular seeds. Seeds are pale yellow and relatively hard. About 90,000-1,00,000 seeds weigh in one kilogram. Seeds can be stored for about a year. Germination is 20-90 percent (Sen Gupta, 1937). Germination is epigeal. It is recommended to soak seeds in water for 12 hours to hasten germination.

Lannea coromandelica (Houtt.) Merr.

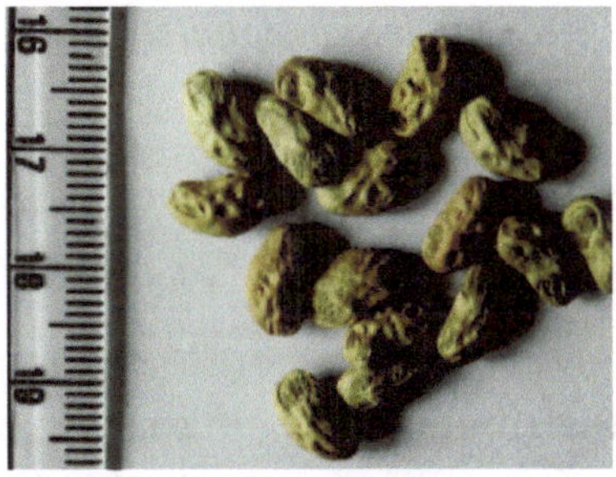

Seeds of Godda

Family : Anacardiaceae.

Synonyms : *Lannea wodier* (Rox.) Adel.

Dialium coromandelicum Houtt.

Common name: Wodier, Godda, Indian ash tree.

Distribution : This is distributed throughout the hotter parts of India. Elsewhere it is found in moist and mixed dry deciduous forests of India.

Habit : Godda is a moderate to large-sized deciduous tree with a spreading crown of 20 m.

Phenology : Flowers appear from March to April. The small green berries are borne in large numbers from May to July.

Uses: The wood is suitable for delicate carving. It is used for posts, light packing cases, pencil and pen holder making, and low-cost quality matches. The leaf is used as fodder and is also tapped for gum known as 'Jhingam' gum which is used in calico-printing, varnishes, etc.

How to identify and Seed Handling: Fruit is a berry. They are reniform or oblong and about 1-1.50 cm long with hard stone. About 7400 to 10600 seed weighs one kg. Germination percentage is 45 to 65 percent. Germination starts within a week and takes about four weeks to complete.

***Leucaena leucocephala* (Lam) de Wit**

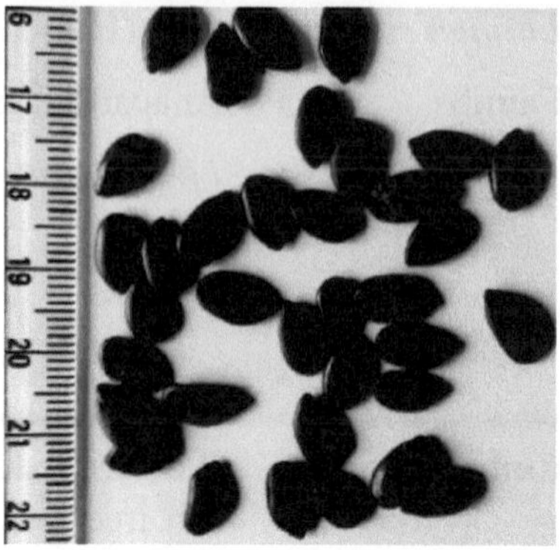

Seeds of Subabul

Family : Fabaceae,

Subfamily : Mimosoideae

Common name: Subabul, River tamarind.

Distribution : It is native to Mexico and grown throughout the world. More than a hundred varieties are known, broadly classified into three types: Hawaiin, Salvador, and Peru (National Academy of Sciences, 1977).

Habit : Subabul is a fast-growing, multipurpose shrub or small evergreen tree growing 2-10 m tall, occasionally reaching 15 m in height.

Phenology : Flowering and fruiting occur almost throughout the year at the same time.

Uses: Leaves are used as fodder to feed the cattle; Flowers yield nectar, and wood has a high calorific value of 4600 kcal kg^{-1} and is used as fuel. Tree yields gum, resin, and tannin.

How to identify and Seed Handling: Fruit is a pod. The pod is oblong and brown. A single seed weighs around 0.059 ± 0.002 g; around 14000 seeds weigh one kg. Seeds are orthodox, having viability of up to five years. Hot water treatment is recommended for better germination, with expected germination up to 45 to 80 percent. Seeds will germinate four days after sowing, epigeal kind of germination.

***Lophopetalum wightianum* Arn.**

Seeds of Balpale

Family : Celastraceae

Synonyms : *Lophopetalum ovatum* Presl. *Euonymus fimbriatus* Baill. ex Laness.

Common name: Balpale, Banate.

Distribution : Tree is indigenous to the Indo-Malayan region and is found in evergreen forests in the Indian Peninsula.

Habit : Balpale is a large evergreen tree with a tall cylindrical stem that reaches 30 m. The stem is usually fluted with 2 to 3 deep flutes.

Phenology : Flowering starts from January to April. Fruiting occurs from May to June. Fruits mature at the commencement of the rainy spell from the end of May to early June.

Uses: Timber is highly used for the manufacture of waterproof plywood and tea-chest plywood. It is also used for building construction, door-frames, railway carriages, cabinet work, pencil making, etc.

How to identify and Seed Handling: Fruit is a three loculus capsule containing about 12 to 14 seeds. Seeds are whitish-cream, broadly winged on all sides, thin and flat. About 4000-5000 seeds weigh one kilogram. Seeds are recalcitrant; dry seeds can store for a week. Since seed viability is very low, sowing freshly collected seeds is advised. Germination is over 90 percent. Germination starts within 2 to 3 days of sowing.

Madhuca longifolia* var. *latifolia

Fruits of Mahua

Family : Sapotaceae

Synonym : *Madhuca latifolia* Macb.

Bassia latifolia Roxb.

Common name: Mahua, Butter tree.

Distribution : The species is found scattered throughout India. Found both in the moist and dry deciduous forest and dry mixed forest.

Habit : Mahua is a medium-sized to a large deciduous tree, attaining a height of 12 to 18 m, with a short bole.

Phenology : Flowering appears from March to May. The fruiting can be seen from July to September. The fruit setting is very low.

Uses: The timber is used for piles, ships, boats, sports goods, furniture, etc. it is also a good fuelwood source. Locally used for agricultural implements. The fleshy corolla is edible. Essential oil is extracted from the flowers. A fatty acid is extracted from the seeds used in soap industries.

How to identify and Seed Handling: Fruit is a green fleshy berry containing 1 to 3 polished brown seeds. Seeds are either double convex or flattened pale brown, and testa is hard. Due to the high oil content in the seeds, seeds cannot be stored for an extended period. About 450 to 500 seeds weigh one kilogram. Fresh seeds have a high germination capacity.

Maesopsis eminii Engl.

Seeds of Maesopsis

Family : Rhamnaceae

Synonyms : *Maesopsis berchemoides* A. Chew

Common name: Musizi.

Distribution : This is naturally distributed in Uganda, Nyanza, Tanzania, Guinea, etc. The tree has been grown in humid conditions in India as a shade tree in the coffee and cardamom plantations.

Habit : Maesopsis is a fast-growing evergreen tree that usually grows up to a height of 15 to 25 m. with a cylindrical bole.

Phenology : Tree flowers twice a year, *i.e.*, March to April and November, and fruits mature from October to January and June to July.

Uses: Timber is one of the most useful light hardwoods used in joinery, indoor construction, etc. It is also used as a potential raw material for manufacturing wooden crates, wooden cases, tool handles, and packaging. The timber is also suitable for the manufacture of plywood.

How to identify and Seed Handling: Fruit is a yellow drupe, turning black on falling. About 750 to 900 seeds weigh one kilogram. Seeds lose viability after few months after collection (maximum six months). Germination starts after 33 to 65 days after sowing, and germination is epigeal. Freshly collected seeds give about 90 percent germination (Mugasha, 1981).

***Melia dubia* Cav.**

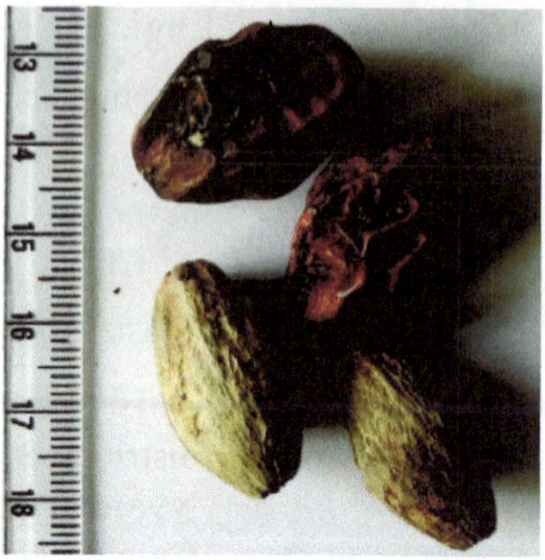

Seeds of Melia

Family : Meliaceae

Synonym : *Melia composita* Willd.

M. robusta Roxb.

M. superba Roxb.

Common name: Malabar neem wood, Hebbevu, Mahanim.

Distribution : It occurs in deciduous forests of Peninsular India, North Bengal, Sikkim, and Bhutan

Habit : This is a fast-growing, large deciduous tree mainly cultivated for timber.

Phenology : Matured fruits will be available from October to February.

Uses: Timber is mainly used in plywood industries for making wallboard, door panel, and furniture. It is also used in agricultural implements, flooring, and boxes.

How to identify and Seed Handling: Fruit is a stony drupe. Each stone contains 1 to 5 black seeds, tiny seeds. It is recorded that one kilogram weighs 300 to 350 stones. Seeds are viable for up to six months. The germination is epigeal. Germination is very low because of hardy stone. Pre-treatments recommended are: splitting the hard endocarp longitudinally, burying seeds in a pit for a year, and soaking in cold water for ten days.

Michelia champaca L.

Seeds of Champaca

Family : Magnoliaceae

Synonym : *Michelia aurantiana* Wall.

Common name: Champaca.

Distribution : This tropical evergreen species is native to India and Indonesia, extending to Taiwan. Primarily found in plains, uplands, & also in forests.

Habit : Champaca is a large evergreen tree with a tapering crown, attaining a height of 30 m.

Phenology : Flowers appear in the hot and rainy seasons, and fruits ripen about August or later.

Uses: Champaca is known worldwide for its large, aromatic yellow blossoms and striking appearance. It is planted along roadsides for its flowers and shadow. Wood is also used in light furniture and indoor works. It is also suitable for moisture-proof plywood.

How to identify and Seed Handling: Fruits are capsules, which dehisce on maturity. After dehiscence, it exposes red-oily arillate seeds that remain attached to the fruit. Seeds are dark brown, polished, and angular; Aril can reduce seed viability by allowing infection by fungus, reducing germination capacity. The seed coat is known to impose and maintain dormancy in several species. About 12000 to 17000 seeds weigh one kilogram (Luna, 1996). Germination is epigeal. Removal of mesocarp and treating seeds with gibberellic acid hasten the germination up to 80 percent.

Mimusops elengi L.

Seeds of Bekul

Family : Sapotaceae

Synonym : *Mimusops parviflora* R. Br.

Common name: Asian bullet wood, Bekul, Eleng.

Distribution : Occurs in the Western Ghats, in the dry evergreen forests of the Eastern Ghats, and also in Andaman, Myanmar, and Sri Lanka.

Habit : Bekul is a slow-growing, large evergreen tree.

Phenology : Matured fruits will be available from June to July.

Uses: The fruits are edible. Wood is used for general construction work.

How to identify and Seed Handling: Fruit is a berry. The fruit is yellowish and ovoid. Seeds are greyish brown ovoid and shining. Around 1500 to 2000 seeds weigh one kilogram. Seeds are recalcitrant and lose viability in a short period after collection. Germination is epigeal. Germination percentage is 30-60 percent. Fresh seeds can give better germination, *i.e.*, 90 percent. Seeds are soaked to smoothen the hard seed coat before sowing.

***Mitragyna parviflora* (Roxb.) Korth.**

Seeds of Mitragyna

Family : Rubiaceae

Synonyms : *Stephegyne parviflora* Korth.

Nauclea parviflora Wills.

Common name: Keli-kadam, Hedu.

Distribution : Tree is a characteristics species of Southern and Northern Indian tropical moist deciduous forest and tropical dry deciduous forest.

Habit : Mitragyna is a medium to large-sized deciduous tree, attaining a height of 15 to 20 m.

Phenology : The flower heads appear from May to July, and the fruit heads become fully formed by October but ripen April-May.

Uses: Timber is usually used for building construction, as planking and rafters, furniture, agriculture implements, etc. Wood is also used as fuel wood. Leaves are used as fodder. The leaves are rich in tannin. The bark yields a cordage fibre.

How to identify and Seed Handling: Fruit is a capsule. Seeds are light, minute, and pointed at either end. The seeds are light brown. About 1000 ripen capsules weigh one kilogram. The germination is 50 to 60 percent (Singh, 1982).

Oroxylum indicum **L. Vent**

Seeds of O. indicum

Family : Bignoniaceae

Common name: Kampong, Sauna

Distribution : This is native to the Indian Subcontinent, found in the Himalayan foothills extending to Bhutan and Southern China, Indochina, and the Malesia regions.

Habit : This is a fast-growing small to medium-sized tree with an open, irregular crown, which can grow up to 10 to 20 m tall.

Phenology : Flowering occurs from May to August. Fruits mature and dehisce, usually from February to May.

Uses: It is used as a green vegetable in Southeast Asia; it is especially liked in Java and is often sold in local markets. The plant is commonly used to supply a range of traditional preparations.

How to identify and Seed Handling: Fruit is a large woody two-halved capsule, having a large number of seeds. Seeds are flat, surrounded by a thin transparent white papery wing. About 10,100 to 12,300 seeds weigh one kilogram. Seeds lose viability after months. Sowing freshly collected seeds is advised for more excellent germination (70 to 90 %). Germination is epigeous.

Peltophorum pterocarpum (DC.) Baker ex Heyne

Seeds of Copper pod

Family : Fabaceae

Synonyms : *Inga pterocarpa DC.* *Peltophorum ferrugineum* Benth.

Subfamily : Caesalpinioideae

Common name: Copper pod, yellow gold mohur.

Distribution : Native to Indo-Malayan region and Northern Australia. In India, it is cultivated mainly in gardens and avenues.

Habit : This is a fast-growing, large evergreen tree with a spreading crown and dense canopy.

Phenology : Pods are available from January to March and September to November.

Uses: This is grown as an ornamental and avenue tree in the garden. Wood is used as fuel.

How to identify and Seed Handling: Fruit is a compressed, indehiscent pod. Seeds are oblong and brown. Around 12500 to 15000 seeds weigh one kilogram. Seeds are usually orthodox and viable for up to one year. Germination is epigeal, and germination is up to 10 percent (Kumar and Bhanja, 1992). Hot water treatment is recommended for better germination.

***Phyllanthus emblica* L.**

Fruits of Amla

Family : Euphorbiaceae

Synonym : *Emblica officinalis* Gaertn.

Cicca emblica L. Kurz

Common name: Indian Gooseberry, Emblica myrobalan, Amla.

Distribution : The plant species is native to India, growing in Sri Lanka, Uzbekistan, South East Asia, and China. They are cultivated widely for fruits.

Habit : Amla is a small to medium-sized deciduous tree, reaching 1 to 8 m in height.

Phenology : Fruiting occurs from October to April.

Uses: The fruits are a rich source of vitamin C. The amla fruit can be eaten raw or cooked in a number of dishes. In traditional Indian medicine, dried and fresh fruits of the plants are used. The whole plant is used in various Ayurvedic medicine herbal preparations, including the fruit, seed, leaves, root, bark, and flowers.

How to identify and Seed Handling: Fruit is a berry with six seeds. Seeds are reniform, trigonous, shiny, and reddish-brown. 65000 to 100000 seeds weigh one kilogram. Seeds are viable for up to three months and fail to germinate after one year (Troup, 1921). Germination is epigeal and up to 40 percent (Luna, 1996). Soaking seeds in hot water for 10 to 12 hours hasten the germination.

Pinus wallichiana Jackson

Seeds of blue pine

Family : Coniferae

Synonyms : *Pinus excelsa* Wall.

Common name: Blue pine, Bhutan pine.

Distribution : This is distributed throughout the temperate Himalayas chiefly at 2000 to 3000 m, but sometimes ascending to 3750 m and descending to 1250 m.

Habit : Blue pine is a large evergreen tree with bluish feathery foliage giving a graceful appearance, attaining a height of 35 m.

Phenology : The flowers appear in April; after pollination scales are closed, fertilization occurs after 1 to 12 months. The cones ripen from September to November.

Uses: Timber is used for planking, door and window frames and shutters, furniture, packing cases, etc. Tree yields resin of better quality than chir pine. Wood is also used as fuelwood and for charcoal.

How to identify and Seed Handling: Fruit is a cone. Scales are arranged in spirals, spathulate, rhomboidal with a dark brown umbo. Seeds are dark brown, ovoid, acute at both ends, and compressed. About 12500 to 22000 seeds weigh one kilogram depending on provenance. Clean, dried seeds can be stored for 14 to 18 months. Germination capacity varies from 60 to 70 percent.

***Pongamia pinnata* (L.) Pierre**

Seeds of Honge

Family : Fabaceae - Faboideae

Synonyms : *Pongamia glabra* Vent.

Derris indica (Lam.) Bennet

Common name: Indian beech, Honge.

Distribution : Honge is found throughout India in plains, along rivers, beach forest of Sunderbans, and cultivated along canal banks, roadsides, and bunds.

Habit : Pongam is a medium-sized, semi-evergreen tree with a spreading shady crowd that can attain a height of 18 m.

Phenology : Flowers appear from April to July, and pods may ripen from February to June.

Uses: Wood is locally used in agricultural implements. Branches lopped for fodder and fuel. Seeds yield non- edible "pongam" oil for soapmaking, bio-diesel, and tanning. Oil cake is good fertilizer. Leaves and roots are used in medicine-one of the well-known tree species for afforestation.

How to identify and Seed Handling: Fruit is a woody, indehiscent pod. Pods are generally one-seeded. Seeds are compressed, reniform, reddish-brown in colour, and wrinkled. 850-900 dry seeds weigh one kilogram (Mathauda, 1955). Seeds are viable for up to one year under ambient conditions. Germination is hypogeal. Soaking seeds before sowing in cold water will improve germination. Germination percentage is 35-70 percent.

Pterocarpus marsupium Roxb.

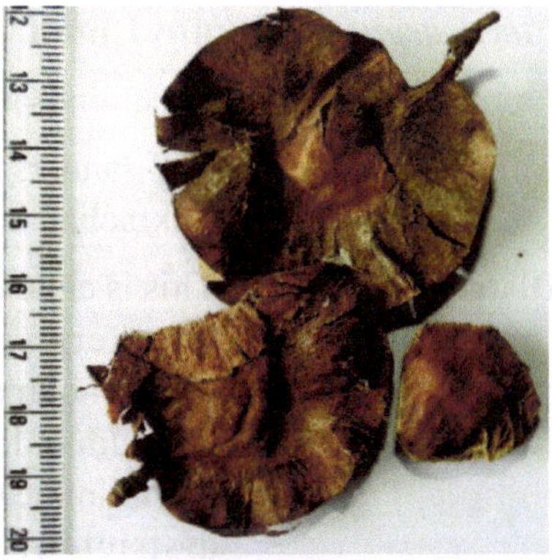

Seeds of Bijsal

Family : Fabaceae

Synonyms : *Lingoum marsupium* (Roxb.) Kuntze

Pterocarpus bilobus Roxb. ex G. Don.

Subfamily : Faboideae

Common name: Bijsal, Indian Kino tree, Honne, Andaman redwood.

Distribution : It is native to Peninsular India, extending from Gujrat to West Bengal. It is found in moist, dry deciduous, and semi-evergreen forests.

Habit : Tall, slow-growing deciduous tree.

Phenology : llow, fragrant flowers appear from August to October, and fruiting occurs from December to March.

Uses: The timber is strong, heavy, hard, and durable. It is used for construction and furniture purposes. Good fodder. Bark, leaves, and flowers have medicinal properties. Kino gum is extracted.

How to identify and Seed Handling: Fruit is a compressed, winged, indehiscent pod. About 1300 to 1400 pods weigh one kilogram. Seeds are orthodox and viable for up to one year. Germination percentage is up to 97 percent and epigeal. Pods being hard, pre-treatment is given to hasten the germination; Cutting the seed wing and soaking them in water; cow dung slurry treatment; placing alternate layers of pods and dead leaves in a pit and flooding it with water.

***Pterocarpus santalinus* Linn.f.**

Fruits of Raktachandana

Family : Leguminosae

Common name: Red sanders, Raktachandana.

Distribution : This is endemic to the Southern Eastern Ghats mountain range of South India. This is found in the Cuddaph district of Andhra Pradesh, Neighbouring areas of Tamil Nadu and Karnataka.

Habit : Red sanders is a small to medium-sized deciduous tree attaining a height of 15 to 8 m.

Phenology : The flowering takes place from April to May. The pods form immediately after flowering but do not ripen until February to March.

Uses: The timber is used to manufacture a particular musical instrument called 'Shamisen.' Wood is used for carving idols. Wood contains Santalin dye. Leaves are used as fodder for cattle.

How to identify and Seed Handling: Pod is winged and one or two-seeded. The dry pods weigh 900 to 1400 to one kilogram. Seeds will be viable for 12 months. Germination percentage varies from 12 to 30% (Rai *et al.*, 1986). Soaking the pods in water for three days and cow dung slurry (48 hours) enhances germination.

Rauwolfia serpentina L. Benth

Seeds of sarpagandha

Family : Apocynaceae

Common name: Sarpagandha, Indian snakeroot

Distribution : It is native to India, Burma, Sri Lanka, China, and Pakistan. In India, it is found in Bihar, Punjab, Orissa, Bengal, Andhra Pradesh, Kerala, and Maharashtra.

Habit : This is an evergreen perennial undershrub.

Phenology : Fruiting occurs from September to November.

Uses: It is a famous ayurvedic herb used for hypertension management. Its root powder (Sarpagandha Churna) is usually better for hypertension than its extract or derivatives to minimize its side effect and increase its effectiveness for long-standing action.

How to identify and Seed Handling: Fruits are fleshy, one to two-seeded. Seeds are white, and 26000 to 35000 seeds weigh one kilogram. Germination is up to 60 percent. Germination starts after 15 to 20 days and lasts up to 40 days. Freshly collected seeds give high germination.

Seeds of raintree

Samanea saman **(Jacq.) Merr.**

Family : Leguminosae - Mimosoideae

Synonyms : *Enterolobium saman* Prain.

Pithecellobium saman Benth.

Common name: Rain tree, Monkey-pod.

Distribution : It is indigenous to Central America, the West Indies, and parts of northern South America. In India, it is widely planted in roadside avenues and pastures for shade.

Habit : The rain tree is a large evergreen tree with a short bole and wide-spreading crown, which attains a height of 18-24 m.

Phenology : Flowers appear in the hot seasons from May to June. Pods form rapidly, but ripening is slow and takes place in March-April of the following year.

Uses: Raintree is a famous avenue tree. Wood is mainly used as fuel wood in urban areas and for petty construction work. It is also used in boat building since the wood is considered durable under water. It is grown as a pasture tree for pods and foliage.

How to identify and Seed Handling: Fruit is a sessile, straight brown-black colored pod. The seed is brownish; 4400 to 7000 seeds weigh one Kilogram. The germination capacity is about 70 to 90 percent. Cold water soaking of seeds can help to hasten germination.

Santalum album L.

Family : Santalaceae

Common name: Sandal, Chandanam.

Distribution : This is native to peninsular India, Indonesia, and Sri Lanka and cultivated in the tropical/ subtropical Asian-Pacific region.

Seeds of Sandal

Habit : Sandal is an evergreen, small-sized semi-parasitic tree also occurs in deciduous forests.

Phenology : The seeds are available twice a year, during May-August and November-December.

Uses: The sandalwood oil extracted from heartwood is used in perfumery, cosmetics and traditional ayurvedic medicine. Wood is used in carving work.

How to identify and Seed Handling: Fruit is a purplish-black drupe with a hard endocarp. Seed weight is 5500- 7000 seeds kg^{-1}. Seeds are intermediate to orthodox, with differing viability from location to location and storage condition. Germination is 8 to 85 per cent (Sen Gupta, 1937; Kumar and Bhanja, 1992). Germination is epigeal and hypogeal. The best-recommended treatments are: 1) Acid scarification with conc. sulphuric acid for 30 minutes. 2) Soaking in 0.05 per cent gibberellic acid for 12 to 16 hours. 3) Seeds are stored for two months to reduce the dormancy; it is recorded that stored seeds will germinate better than fresh seeds (Srimathy and Kulkarni 1995; Luna, 1996).

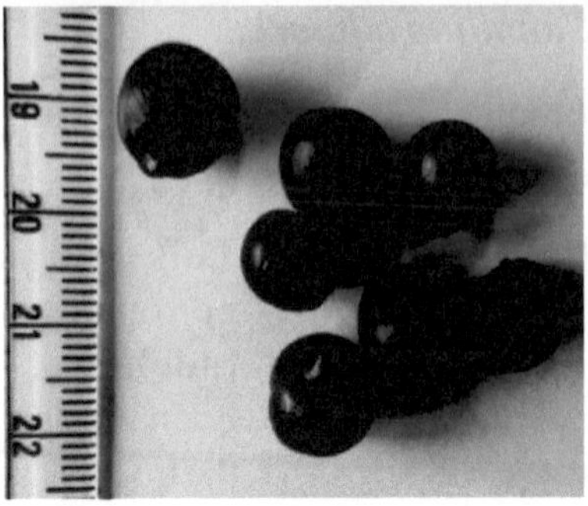

Seeds of Soapnut

Sapindus emarginatus **Vahl.**

Family : Sapindaceae

Synonyms : *Sapindus trifoliatus* Linn.

Common name: Soapnut.

Distribution : Tree is native to South India and is found in deciduous forests; it constitutes Southern Tropical Wet evergreen and dry evergreen forests.

Habit : Medium-sized deciduous trees with a spreading crown, attaining 18 m.

Phenology : Tree flowers from October to December. Fruits ripen in the hot seasons from February to April.

Uses: Timber is locally used for carts, agricultural implements. It is also used as fuelwood. Fruit is subtitle for soap in washing silk, woollen and delicate articles. Fruits and root bark are used as a fish poison.

How to identify and Seed Handling: Fruit are fleshy, 2 or 3 lobed. Seeds blackish, smooth, very hard. The seeds are viable for up to 2 years. Seed weight varies between 2000 to 2500 per kilogram. The germinative capacity is 65 to 80 per cent. Sandbag treatment with alternative wetting and drying fastens germination by breaking hard seed coat.

Sapindus mukorossi Gaertn.

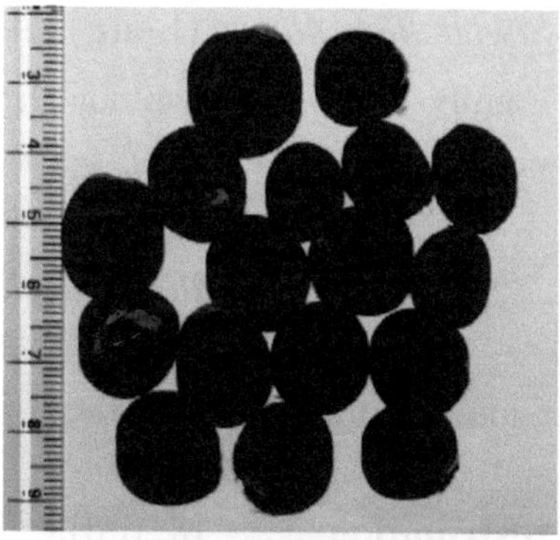

Seeds of Reetha

Family : Sapindaceae

Synonym : *Sapindus detergens* Roxb.

Common name: Reetha, Washnut

Distribution : The species is widely grown in the Indo-Gangetic plains' upper reaches, Shivaliks and sub-Himalayan tracts at altitudes from 200 m to 1500 m. Possibly indigenous to Western Himalaya.

Habit : It is a fairly large, deciduous tree, usually up to 12 m in height, sometimes attaining 20 m

Phenology : It flowers during summer. The flowers are small and greenish-white. The fruit appears in May-July and ripens by November-December.

Uses: The wood is utilized for rural building construction, oil and sugar presses, and agricultural implements. *Sapindus mukorossi* tree has the highest saponin content. Saponin is a natural detergent commonly used for cleaning, among many other things. Soapnuts have been used medically.

How to identify and Seed Handling: Fruits are solitary globose, round nuts 2 to 2.5 cm diameter, fleshy, saponaceous and yellowish-brown in colour. The seed is bounded in a black, smooth and hard globose endocarp. 750 to 900 seeds weigh one kilogram. Seeds are viable for one year. Germination is epigeal. To ensure the best germination results, the seed is soaked in warm water for at least 24 hours and then sown.

Schleichera oleosa (Lour.) Oken

Seeds of Schleichera

Family : Sapindaceae

Synonym : *Schleichera trijuga* Willd.

Pistacia oleosa Lour.

Common name: Lactree, Ceylon, Chakote.

Distribution : In India, it occurs sporadically on the low hills of the Himalayas and in semi-evergreen and moist deciduous forests. It is also found in Myanmar, Sri Lanka, and Indonesia.

Habit : Slow growing, medium-sized deciduous tree attaining a height of 20 m.

Phenology : Fruits will mature in March-June when fruits turn brownish-green and the pulp is yellow.

Uses: The ripe fruit is eaten raw. Seed are the primary source of Kusum oil. Foliage is used as fodder to feed livestock and cultivated as a host plant in lac production. Wood is locally used in construction and tool handles.

How to identify and Seed Handling: Fruits are simple, fleshy and of the Berry category, either one seeded or two-seeded. Seed is irregular or ellipsoidal, slightly compressed and has a thick brown seed coat on its surface. About 1500 to 2200 seeds weight one kilogram. Seeds are orthodox and viable for up to two years. Germination is epigeal and 50 to 80 per cent (Luna, 1996). Soaking seeds in hot water fasten the germination.

Semicarpus anacardium L.f.

Fruits of Kadugeru

Family : Anacardiaceae

Common name: Oriental cashew, Marking nut tree, Dhobi nut.

Distribution : Africa, Indo-Malesia, China and Australia. Distributed throughout India except for Jammu and Kashmir, Punjab, Haryana, Sikkim, Arunachal Pradesh.

Habit : *Semicarpus anacardium* is a moderately sized deciduous tree, reaching a height of 12-15 m.

Phenology : Fruiting occurs in December to January. Fruits remain on the tree from February to June.

Uses: The fruits and their oil (28-36%) have medicinal property. Juiceof the fruit pericarp is used for marking cotton clothes. Wood is suitable for packing cases.

How to identify and Seed Handling: Fruit is a drupe of 2 to 2.5 cm long, obliquely ovoid, smooth and shining, black when ripe, borne on the orange-coloured receptacle. 460 to 800 seeds weight one kilogram. The seed viability is six months. Germination is epigeal and up to 21 percent (Sen Gupta, 1837). Alternate days soaking and drying seeds in cow dung for 15 days can hasten seed germination.

Seeds of Kadambate

***Spondias pinnata* (L.f.) Kurz**

Family : Anacardiaceae

Synonym : *Mangifera pinnata* L.f.

Spondias mangifera Willd.

Common name: Indian hof-plum

Distribution : It is widely occurring in India. It has been cultivated in Sri Lanka and Malaysia for edible purpose. It naturally found in Moist and Dry deciduous forests.

Habit : Kadambate is a medium-sized deciduous tree attaining height up to 30 m.

Phenology : The matured fruits will be available from December to May (Sen Gupta, 1937)

Uses: The fruits are used in making pickles and edible. The bark, leaves and roots have medicinal value. Timber is lightweight and used for making package cases and matchboxes.

How to identify and Seed Handling: Fruit is a fleshy drupe, turn yellow on maturity. Seeds are separated by depulping. The stone is fibrous and woody. Around 300-320 stones will weigh one kilogram. Seeds are stored for the short period since it goes on losing viability. Germination percentage is 60-80 per cent. Germination is Epigeal. Stones are clipped before sowing for hastening the germination.

Strychnos nux-vomica L.

Seeds of snakewood

Family : Longaniaceae

Common name: Nux vomica tree, Crow fig, Snakewood, Kanjira.

Distribution : The species is native to India. Commonly found in the moist and dry deciduous type of forest.

Habit : It is a slow-growing deciduous tree, grows up to 35 m.

Phenology : Fruits will mature from December to June.

Uses: The seeds are having alkaloids, which are used in medicines. Wood is moderately hard and used in agricultural implements, axe hammers and ornamental panels.

How to identify and Seed Handling: Seeds are discoid, compressed with concave and convex sides with fine silky hairs. On an average 450 to 600 seeds weigh one kilogram. Seeds are orthodox and viable for one year. Germination is epigeal with 15- 52 per cent. Seeds are soaked in boiling water followed by soaking in cold water or seeds soaked in cow dung slurry for 24 hours will give better germination.

Swietenia mahagoni (L.) Jacq.

Seeds of S. mahagoni

Family : Meliaceae

Synonyms : *Swietenia fabrilis* Salisbury
Cedrus mahogany (L.) Miller

Common name: Cuban mahagany.

Distribution : *S. mahagoni* is a humid zone species with natural distribution in the West Indies. It has been extensively planted mainly in southern Asia (India, Sri Lanka, and Bangladesh) and the Pacific region.

Habit : Mahagany is a fast-growing evergreen tree, attain height up to 30-35 m.

Phenology : The flowers appear from April to May. The pods ripen from October to December.

Uses: Mahogani has potential use for timber production plantations, mainly in furniture. It is also planted as an ornamental tree in avenues.

How to identify and Seed Handling: Fruits are woody capsules having five valves that are dehiscent after maturity. Seeds are attached to the inner woody columella by hanging pendulous to the apex by wings. Winged seeds brown in colour, compressed. 1800 to 2200 seeds weigh one kilogram (Dent, 1948). Seeds that are orthodox and viable can be preserved for up to six months. The germination percentage of fresh seeds is 60-90 per cent. Seed germination may be enhanced by soaking in water for 12 hours. Germination is hypogenous.

***Tamarindus indica* L.**

Seeds of *T. indica*

Family : Fabaceae

Subfamily : Caesalpinioideae

Synonyms : *Tamarindus occidentalis* Gaertn.

T. umbrosa Salisb.

Common name: Tamarind, Indian tamarind

Distribution : This is indigenous to dry savannahs of tropical Africa. It was introduced to Asia, where it has become naturalised.

Habit : It is an evergreen, slow-growing tree reaching up to 30 m tall with a dense, spreading crown and short trunk.

Phenology : Tree flowers from April to June, the pods ripen from February to April of the following year.

Uses: Tamarind is mainly grown for its fruits but also a valuable timber species. Seeds are also edible and used in a variety of dishes. Wood is good firewood and is excellent for making charcoal.

How to identify and Seed Handling: Fruits are indehiscent pod, brittle, more or less curved and constricted between the seeds. Seeds embedded in the sticky pulp. Seeds with hard and smooth testa are reddish, dark brown or shiny black. There are 900-2000 seeds kg^{-1} (Singh *et al.*, 2017). The seeds are orthodox and can be stored at low moisture content. If stored at 5-10 °C, the seeds can retain high viability for several years without pre-treatment. Germination percentage varies between 55 to 70 per cent.

***Tectona grandis* L.f.**

Seeds of Teak

Family : Lamiaceae

Common name: Teak, Teck, Tega

Distribution : Teak is naturally distributed in the Indian Subcontinent, Myanmar, Thailand, and Laos's western parts.

Habit : It is a moderately growing large deciduous tree. It occurs in Tropical moist and Dry deciduous forest types of central and southern India.

Phenology : Flowering appears after the rainy season *i.e.*, June to September and fruits will be available from November to January.

Uses: Teak is highly durable, the most important timber among timber yielding trees. They are extensively used for shipbuilding, bridging, cabinet works, railways carriage, decorative panelling, general carpentry, etc. It is used for making excellent plywood.

How to identify and Seed Handling: Fruit is a hard nut covered with calyx and mesocarp. Each nut is containing 1 to 4 seeds. 1500 to 3000 nuts weigh one kilogram. Seeds are orthodox, viable for more than one year. Germination is epigeal. Germination percentage is 10 to 70 per cent based on the source. Weathered seeds perform well (alternate wetting and drying). The other methods followed are boiling water treatment, scorching fire, acid treatment, cow-dung treatment, etc.

Terminalia arjuna Bedd.

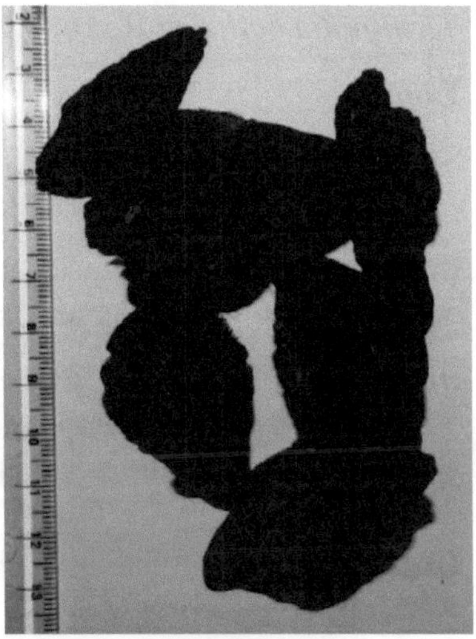
Fruits of Arjuna

Family : Combretaceae

Synonym : *Terminalia glabra* W. et *Pentaptera arjuna* Roxb.

Common name: Arjuna.

Distribution : It is generally distributed in India, all along the forest streams, ravines, river banks.

Habit : Arjuna is a large, evergreen tree with drooping branches, attains a height of 24 m.

Phenology : The flower spikes appear in April to July. The fruits mature from February to May, after a year of flowering.

Uses: The wood is used for carts, agricultural implements, boat building, mine props, etc. It is suitable for the manufacture of plywood of second grade and tea-chest. Wood is also used as fuelwood. The fruits contain 7 to 20 per cent tannin.

How to identify and Seed Handling: Fruits ovoid oblong with 5 to 7 raised woody ribs (wings), single-seeded, hard pericarp acts as a physico-chemical barrier for water penetration. About 175 to 450 fruits weigh one kilogram. Seeds soaked in hot water for 75-96 hours give 90 per cent germination. Treating seeds in cow dung slurry will hasten the germination.

***Terminalia bellirica* Roxb.**

Seeds of Thare

Family : Combretaceae.

Synonym : *Myrobalanus bellirica* Gaertn.

Common name: Belleric myrobalan, Behera, Thare.

Distribution : The species is commonly found in India except in arid regions. It occurs mainly in moist and dry deciduous and semi-evergreen forests.

Habit : This is a large, handsome and moderately fast-growing deciduous tree.

Phenology : Flowering starts from March to May, and fruiting is almost ripe from November to February.

Uses: Wood is hard but not durable. It is used in packing cases, tea chests, plywood, blackboards. Fruits are used in medicine and astringent. Leaves are used as fodder and used to feed tassar silkworms.

How to identify and Seed Handling: Fruit is a subglobose drupe. Around 350 to 400 stones weigh one kilogram. Stones are light yellow coloured. The viable is marinated up to one year under average condition. Germination is hypogeal. Germination percentage is up to 70 per cent. To soften the hard seed coat, seeds are soaked in cold water for 24 hours to several days.

Terminalia chebula (Gaertn.) Retz.

Fruits of chebula

Family : Combretaceae

Synonym : *Myrobalanus chebula* Gaertn.

Terminalia parviflora Thwaites

Common name: Black yrobalan, Chebulic myrobalan. Gall nut.

Distribution : It occurs throughout India, West Bengal, Assam, Sri Lanka and Myanmar.

Habit : Gall nut tree is a slow-growing moderate deciduous tree.

Phenology : Flowers appear from April to June. The fruits ripen from October to March.

Uses: The timber is used for construction, as posts and beams, carts, tool handle, etc. Fruits are used in Triphala choorna, an Indian system of medicine (Chebulic myrobalan).

How to identify and Seed Handling: Fruit is a woody drupe and about 190-200 dried fruits weigh one kilogram. Seeds are orthodox and stored for up to one year with retaining viability. Germination is up to 60 percent(Sen Gupta, 1937). Germination is epigeal. For better germination, seeds are treated by fermentation for 15 to 20 days, or seeds may be clipped at their broad ends and soaked in cold water for 24 hours.

***Terminalia crenulata* Roth.**

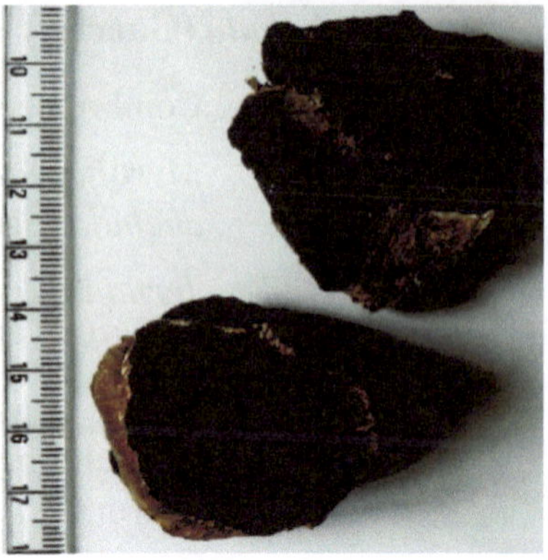

Fruits of Karimatti

Family : Combretaceae

Synonym : *Terminalia tomentosa* Heyne ex Roth.

Terminalia tomntosa Roth. Cl.

Common name: Indian laurel, Karimatti.

Distribution : This is widely distributed species in South-west India and common in the Western Ghats. Gregariously occurs in the tropical Semi-evergreen, moist, and dry tropical forests.

Habit : Karimatti is a moderately fast-growing, large deciduous tree.

Phenology : Flowering appears from May to June. The fruits formed soon after the end of the flowering season, mature in March to April.

Uses: The wood of karimatti is used for agricultural implements, building purpose, fuelwood and charcoal production.

How to identify and Seed Handling: Fruit is a drupe with four to five wings; each contains pale yellowed coloured seed. 520 to 550 fruits weigh one kilogram. Seeds are intermediated and can be stored for one year under sealed conditions. Germination is epigeal. Germination percentage is up to 72 per cent (Dent, 1948). The dewinged seeds are soaked in water or cow dung for better and faster germination.

Toona ciliata Roem.

Seeds of Toon

Family : Meliaceae

Synonym : *Cedrela toona* Roxb. *ex* Willd

Common name: Red cedar, Toon

Distribution : Toon is widely distributed in the Sub-Himalayan tract and valleys of outer Himalaya, Bihar, Assam, Manipur, Western and Eastern Ghat, it is abundantly found in semi-evergreen and evergreen forests.

Habit : Toon is a large semi-deciduous tree, attain 20 to 30 m in height.

Phenology : Flowers appear during November- January and March- April based on the localities. The seeds will be released from the capsules at various intervals.

Uses: The wood is used in structure and furniture and boats, musical instruments, carts, etc. Leaves are used as fodder and dye can be extracted from the flowers. Bark has medicinal property.

How to identify and Seed Handling: Fruits are woody capsules with winged seeds that dehisce after maturity. Seed is very light; Seed weight is 220000-250000 seeds kg^{-1} (TROUP, 1921). Seeds have less viability and can be stored for one month in ambient condition. Germination capacity is 60-80 per cent and germination is epigeal. Sowing of fresh seeds is recommended for better germination and no pre-sowing treatment is required.

Wrightia tinctoria (Roxb.) R. Br.

Seeds of Beppale

Family : Apocynaceae

Synonym : *Nerium tinctorium* Roxb.

Common name: Pala indigo, Beppale.

Distribution : This is commonly found in peninsular India and distributed mainly in moist and deciduous forests.

Habit : Beppale is a slow-growing, small deciduous tree.

Phenology : Seeds will mature from January to February.

Uses: The wood is mainly used in making toys (Chanpatna toys), carvings, wooden combs, bedsteads logs, Matchboxes and printing blocks.

How to identify and Seed Handling: Fruits are paired oblong follicles. Follicles are slender and pendulous with many oblong, pale yellow-brown coloured seeds. 22000 to 22500 seeds weigh a kilogram. Seeds are soaked in cold water for better germination (40 to 55%).

3

References

Adeniji, I.T., Adio, A.F., Iroko, O.A., Kareem, A.A., Jegede, O.C., Kazeem-Ibrahim, F., Adewole, T.O. and Adeosun, A.O. (2014). Pre-treatment of seeds of *Annona squamosal* (Sugar apple) a non-timber forest product. Res. Plant Sci, 2: 50-52.

Anon (1897). Nursery Manual. National Wasteland Development Board, New Delhi.

Anonymous (1950). Annual Report of the Silvicultural Research in Bombay State for the year 1948-49.

Anonymous (1962). Silvicultural Research in Rajasthan: A report.

Awang, K. and Taylor, D. (1993). *Ac*acia mangium; Growing and utilization. MPTS monograph series No. 3 (No. PB-95-160032/XAB), Winrock International Inst. For Agricultural Development, Arlington, VA (United States).

Carlowitz, P.G. (1991). Multipurpose trees and shrubs; sources of seeds and inoculants. International Council for Research in Agroforestry (ICRAF).

CSIR (1986). The Useful Plants of India. Publications and Information Directorate, CSIR, Delhi.

Dent, T.V. (1948). Seed storage with particular reference to the storage of seed of Indian forest plants. Manager. Government of India Press.

Ghosh, R.C. (1977). Handbook on afforestation techniques.

Gupta, R.K. (1993). Multipurpose trees for agroforestry and wasteland utilization. Oxford & IBH Publishing CO.

Kala, A., Raja, P., Soosairaj, S. Balagury, B. (2014). Reproductive biology and seed germination of tropical evergreen tree *Canarium strictum* Roxb. Journal of Biodiversity and Environmental Science, 5(6): 24-32.

Khanpara, K., Renuka, V. and Harisha, C.R. (2012). A detailed investigation on Shikakai (Acacia concinna Linn.) fruit. Journal of Curent Pharmaceutical Research, 9(10): 06-10.

Kindt, R., Muasya, S., Kimotho, J. and Waruhiu, A. (1997). Tree seed suppliers directory: sources of seeds and microsymbionts (Vol.634, No. T786). ICRAF, Nairobi (Kenia).

Kumar, S.V. and Bhanja, M. (1992). Forestry seed manual of Andhra Pradesh. Hyderbad: Research & Development Circle, Andhra Pradesh Forest Department, p. 100.

Luna, R.K. (1996). Plantation trees. International Book Distributors.

Mathauda, G.S. (1955). *Pongamia pinnata* (L.) Pierre Fam. Leguminosae-Papilioneae. Indian Forester. 81(9): 563-565.

Mugasha, A.G. (1981). The Silviculture of Tanzanian indigenous tree species. II. *Maesopsis eminii.* Tanzania Silviculture Technical Note No. 52.

Natarajan, N. and Ravi, R.S.V. (1988). Studies of maximisation of germination in *Acacia auriculiformis*. Tropics. National Academy of Sciences, Washington DC, USA.

Rai, S.N. (1999). Nursery and planting techniques of forest trees in tropical South-Asia. Punarvasu Publications.

Rai, S.N., Nagaveni, H.C. and Padmanabha, H.S.A. (1986). Germination and viability of some tree seeds. Van vigyan, 24(1-2): 8-12.

Sen Gupta, J.N. (1939). *Dipterocarpus* (gurjan) forests in India and their regeneration. Indian Forest Record, 3.

Sen Gupta, J.S. (1937). Seed weights, Plant percents, etc., for forest plants in India. Silviculture 2(3): 175-222, Forest Research Institute.

Shetty, B. (1972). Cultivation of cashew in the Forest lands of Tamil Nadu. Proceedings of the Symposium on Man-made Forests. Society of Indian Foresters, Dehra Dun.

Singh, R.V. (1982). Fodder trees of India. *Fodder trees of India.*

Singh, T.J. and Gupta, T. (2017). Stduies of the Reproductive Biology of *Toona ciliata* M. Roem under the Himachal Pradesh Condition. International Journal of Bio-Resource and Stress Management, 8(3): 437-443.

Srimathi, R.A. and Kulkarni, H.D. (1995). Recent advances in research and management of Sandal (*Santalum album* L.).

Subramanian, K.N., Bedell, P.E., Gurumurti, K., Geroge, M., Mandal, A.K., Pillai, S.M. and Singh, B.G. (1992). Casuarinas: trees of multiple utility. Casuarinas; trees of multiple utility. ICRFI.

Tewari, D.N. (1992). A monograph on teak (*Tectona grandis* Linn.f.). International book distributors.

Tompsett, P.B. and Kemp, R. (1996). Database of tropical tree seed research; special reference to the Dipterocarpaceae, Meliaceae and Araucariaceae. V.01; V.02; User manual (No. 634.987 D232). Royal Botanic Gardens, Kew Richmond (RU).

Troup, R.S. (1921). The Silviculture of Indian Trees, Vol. I, II, III, Clarendon Press, Oxford

Appendix

List of species shown in the cover page

Scientific name of species with vernacular name

1. *Lophopetalum wightianum* (Balpale, Banate) **2.** *Dipterocarpus indicus* (Gurjan, Kalpine) **3.** *Pongamia pinnata* (Honge) **4.** *Mimusops elengi* (Bekul, Eleng, Ranja) **5.** *Butea monosperma* (Palas. Muttuga) **6.** *Ensete superbum* (Rock Banana, Kallubaale) **7.** *Adenanthera pavonina* (Manjatti) **8.** *Strychnos nux-vomica* (Kaasarakaana) **9.** *Pinus wallichiana* (Pine) **10.** *Artocarpus hirsutus* (Hebbalasu) **11.** *Eucalyptus tereticornis* (Nilgiri) **12.** *Tamarindus indica* (Hunase) **13.** *Alnus nepalensis* **14.** *Elaeocarpus munroii* (Kal bikki) **15.** *Cassia fistula* (Kakke) **16.** *Caesalpinia mimosoides* **17.** *Swietenia mahagoni* (Mahagani) **18.** *Gnetum ula* (Beethu, Navuru Katte)

Fruits and Flowers of some locally available tree species

Artocarpus hirsutus

Santalum album

Chrysophyllum roxburghii

Syzygium cumini

Aquilaria agallocha

Gnetum ula

Melia dubia

Terminalia paniculata

Jacaranda mimosifolia

Acrocarpus fraxinifolius

Bixa orellena

Cassia fistula

Bombax malabarica

Mimusops elengi

Michelia champaca

Adenanthera pavonina

Coscinium fenestratum

Canthium dicoccum

Tectona grandis

Terminalia bellirica

Ceiba pentandra

Rauwolfia serpentina

Helicteres isora

Terminalia chebula

Mellotus tetracoccus

Ziziphus oenoplia

Ziziphus rugosa

Acacia auriculifomis

Acacia mangium

Memecylon indica

Garcinia indica

Careya arborea

Myristica malabarica

Aphanamixis polystachya

Randia spinosa

Holoptelea integrifolia

Anogeisus latifolia

Leucaena leucocephala

Mallotus philippensis

Dalbergia sissoo

Sapindus emarginatus

Chukrasia tabularis

Artocarpus altilis

Abrus precatorius

Diospyros discolor

Myristica fragrans

Flacourtia montana

Duabanga grandiflora

Couroupita guianensis